# ENDS OF THE EARTH

**ALSO BY NEIL SHUBIN**

*Some Assembly Required*
*The Universe Within*
*Your Inner Fish*

DUTTON

# ENDS OF THE EARTH

Journeys to the Polar Regions in Search of Life, the Cosmos, and Our Future

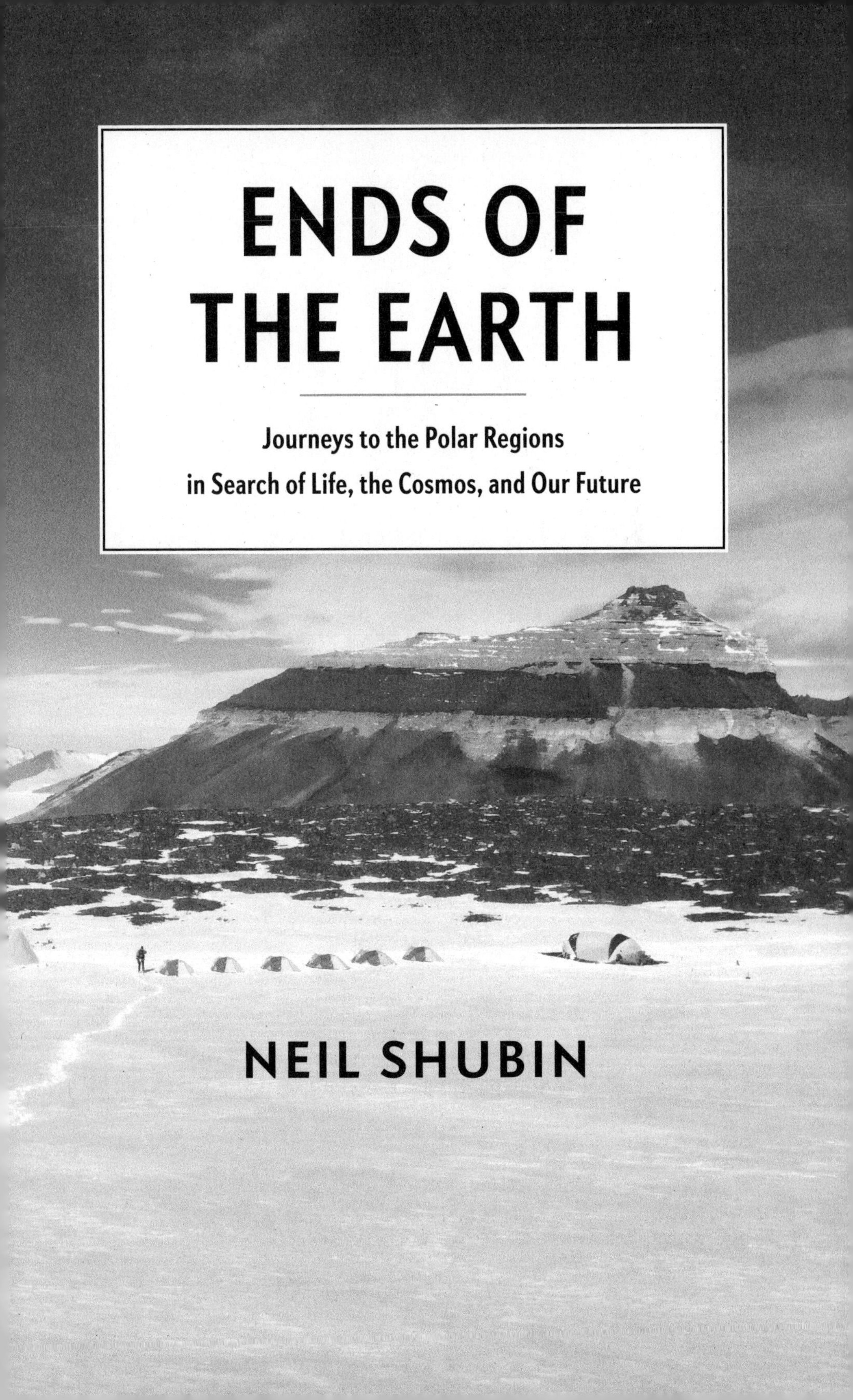

NEIL SHUBIN

DUTTON
An imprint of Penguin Random House LLC
1745 Broadway, New York, NY 10019
penguinrandomhouse.com



Book design by Nancy Resnick

LIBRARY OF CONGRESS CATALOGING-IN-PUBLICATION DATA

Names: Shubin, Neil, author.
Title: Ends of the Earth: journeys to the Polar Regions in search of life, the cosmos, and our future / Neil Shubin.
Description: New York: Dutton, [2025] | Includes bibliographical references and index.
Identifiers: LCCN 2024031260 | ISBN 9780593186527 (hardcover) | ISBN 9780593186534 (ebook)
Subjects: LCSH: Polar regions—Discovery and exploration. | Polar regions—Environmental conditions. | Shubin, Neil—Travel—Polar regions. Classification: LCC G587 .S55 2025 | DDC 910.911—dc23/eng/20241121
LC record available at https://lccn.loc.gov/2024031260

Printed in the United States of America

1 3 5 7 9 10 8 6 4 2

The authorized representative in the EU for product safety and compliance is Penguin Random House Ireland, Morrison Chambers, 32 Nassau Street, Dublin D02 YH68, Ireland, https://eu-contact.penguin.ie.

To my colleagues on polar expeditions:
remembering the times we shared and all we learned

# CONTENTS

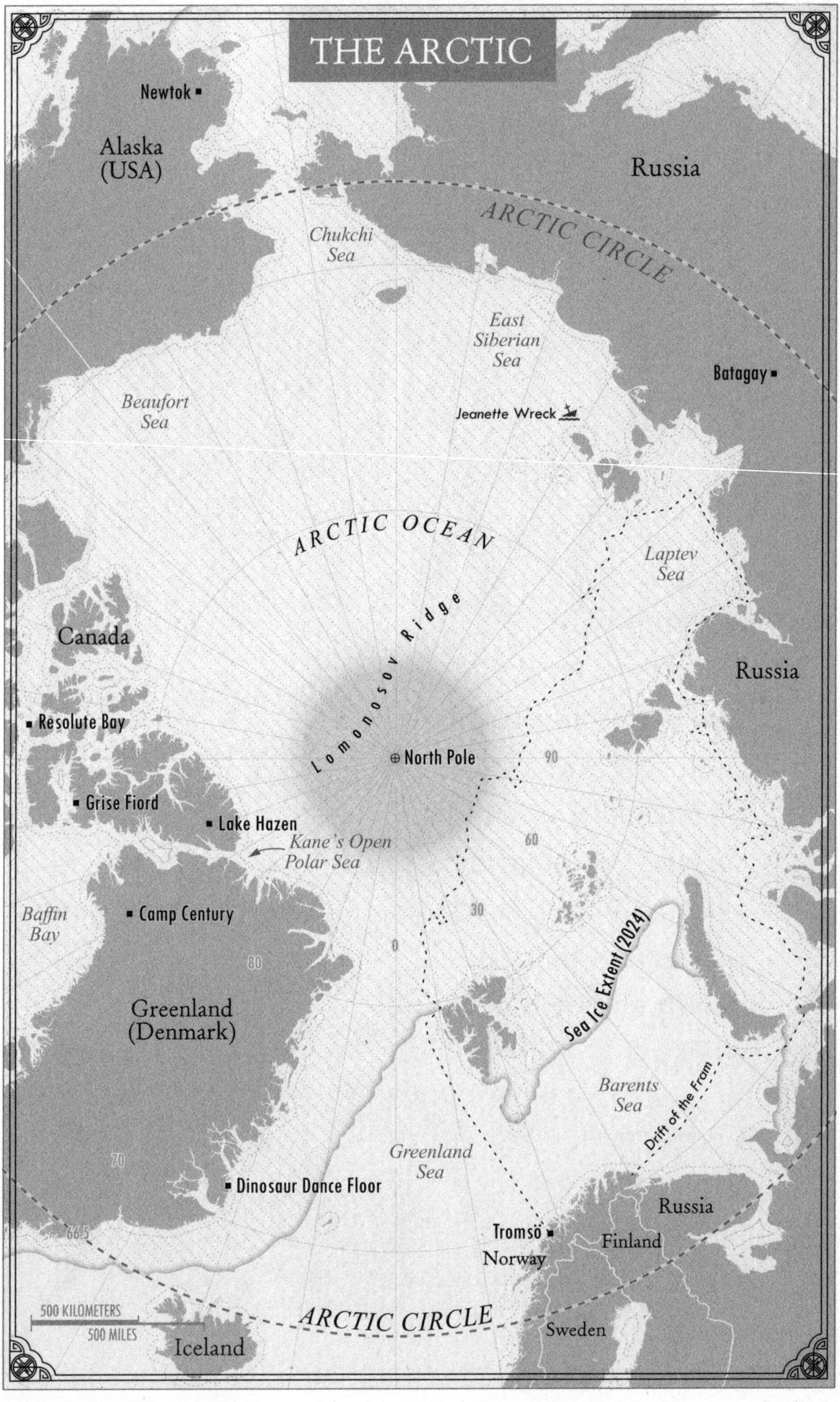
THE ARCTIC
Newtok
Alaska (USA)
Russia
ARCTIC CIRCLE
Chukchi Sea
East Siberian Sea
Batagay
Beaufort Sea
Jeanette Wreck
ARCTIC OCEAN
Laptev Sea
Lomonosov Ridge
Canada
Russia
Resolute Bay
North Pole
90
Grise Fiord
Lake Hazen
60
Kane's Open Polar Sea
Baffin Bay
Camp Century
30
0
80
Sea Ice Extent (2024)
Greenland (Denmark)
Barents Sea
Drift of the Fram
Greenland Sea
70
Dinosaur Dance Floor
Russia
66.5
Tromsö
Finland
Norway
500 KILOMETERS
500 MILES
ARCTIC CIRCLE
Sweden
Iceland

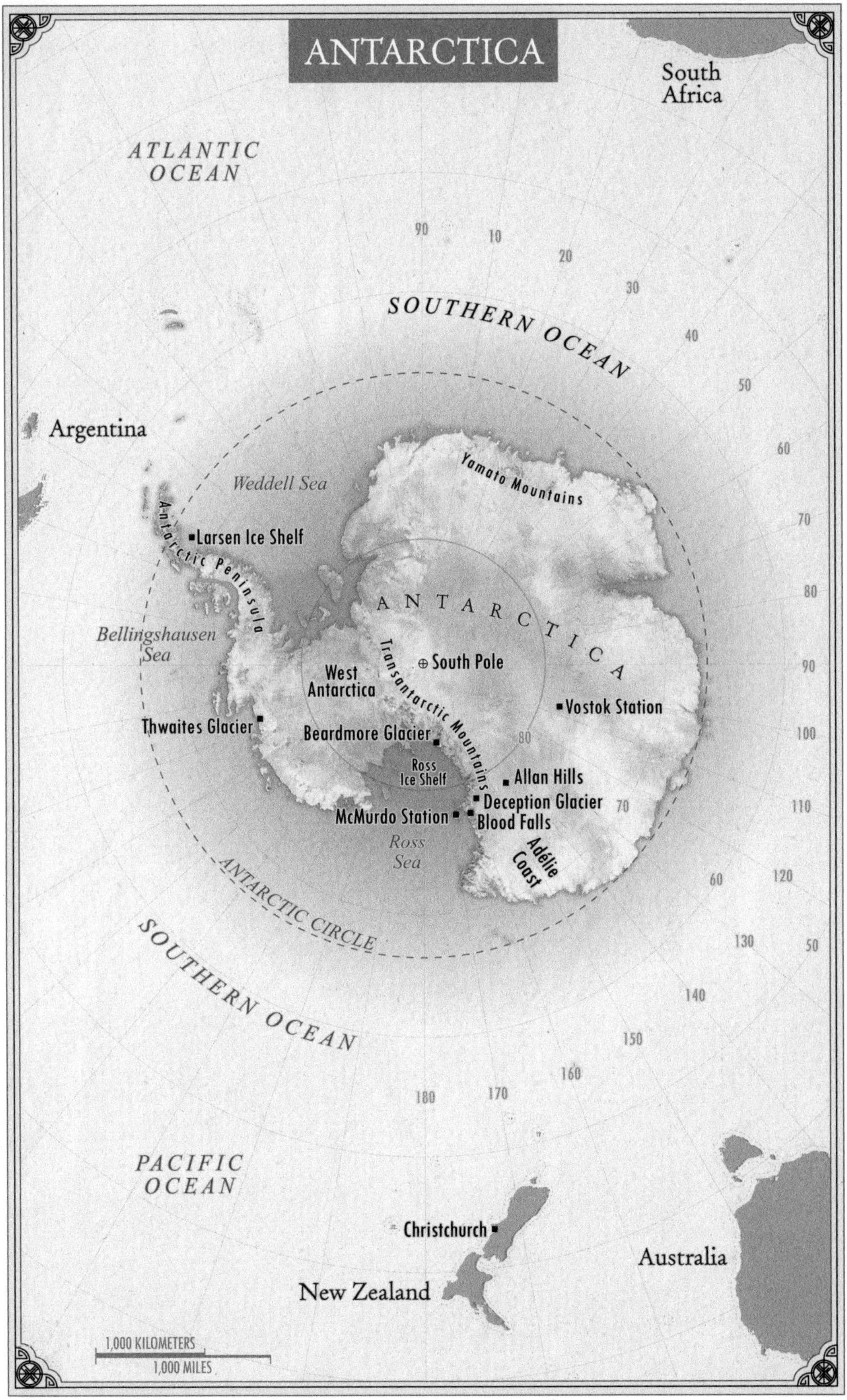
ANTARCTICA
South Africa
ATLANTIC OCEAN
90
10
20
30
40
50
60
70
80
90
100
110
120
130
140
150
160
170
180
SOUTHERN OCEAN
Argentina
Yamato Mountains
Weddell Sea
Antarctic Peninsula
Larsen Ice Shelf
ANTARCTICA
Bellingshausen Sea
West Antarctica
South Pole
Transantarctic Mountains
Vostok Station
Thwaites Glacier
Beardmore Glacier
80
Ross Ice Shelf
Allan Hills
Deception Glacier
70
McMurdo Station
Blood Falls
Ross Sea
Adélie Coast
60
50
ANTARCTIC CIRCLE
SOUTHERN OCEAN
PACIFIC OCEAN
Christchurch
Australia
New Zealand
1,000 KILOMETERS
1,000 MILES

# ENDS OF THE EARTH

# PROLOGUE

Hunched inside a small propeller plane with my knees pressed against a pallet of tents, food, and fossil-collecting gear, I felt I could almost touch the icebergs below as the coast of Greenland emerged in the distance. It was July of 1988, and I was the lone student—a novice in camping and fieldwork—teamed up with three veteran field paleontologists on a hunt for early dinosaurs in East Greenland. Our plane began the flight at a massive airstrip at a naval air station outside of Reykjavík, where it was dwarfed by aircraft taking off to monitor Soviet submarine activity in the area. Designed to land on unpaved landing strips or patches of tundra in remote Arctic valleys, the plane flew so slowly that I feared it might fall out of the sky.

Once over our Greenlandic target, the pilot was hard at work looking for a safe landing spot amid tundra and rock. He banked the plane in a valley bordered by ice-capped red and green mesas and buttes, hunting for a 1,500-foot-long area free of rocks where he could set down. As the plane made steep turns, the cliffs ahead filled the windshield. Seeing our pilot's mane of gray hair brought a strange sense of comfort as I clung to both my seat and the adage "There are old bush pilots and there are bold ones." After three or four passes, he performed mock landings by touching tires to tundra to test the stability of the ground. Mud, large hummocks of tundra, and boulders are the bane of bush landings in the Arctic. Eventually satisfied with his choice, he brought us to the ground, with bumps and shimmies that shook items loose around the cabin.

The pilot had to depart quickly because of the threat of bad weather on his return, so we unloaded our gear in an impromptu bucket brigade. I was already covered with mud while our pilot, who did more than his fair share of lugging, stood pressed and clean. The roar of the plane on his departure marked the beginning of our new life for the next six weeks. Rocks, ice, and polar wildlife—along with the research we were there to do—were now our entire world.

Science in the polar regions means confronting emotional, physical, and logistic challenges in order to ask fundamental questions about the past, present, and future of life and our planet. Since I've proceeded from that rookie experience in Greenland to lead my own expeditions in the Arctic and Antarctica over the past three decades, my original trepidation has been replaced with gratitude for the privilege of working in some of the most extraordinary landscapes on Earth. How many times can you work in places that few, if any, humans have ever encountered before? And to do that exploration in the service of understanding the fundamentals of our world and the ways it came to be?

For European cultures, polar science began as a nationalistic enterprise, driven by attempts to achieve or exploit the points farthest north or south. The race for the poles, or for the discovery of a Northwest Passage connecting the Atlantic and Pacific Oceans, pitted nations against one another in often tragic attempts to be the first to arrive at a certain destination on the globe. Scientific discoveries emerged as teams mapped these regions and studied the natural history, peoples, and climate of the new terrain. Despite competition, much of our current scientific understanding of polar regions ultimately arose from international collaborations that were unique in their breadth and altruism.

Starting in 1870, scientists declared occasional "International Polar Years," when researchers from different countries would form collaborations on geological and meteorological studies at the poles. During the International Polar Year of 1957–1958, scientists used modified rock-drilling equipment to core into the ice of Greenland and Antarctica. This effort brought the leaders of many different countries to the bargaining table, where they went on to craft an extraordinarily broad and forward-thinking international agreement. Stating that the sole purpose of any nation's use of Antarctica is for scientific discovery alone, the Antarctic Treaty was signed by twelve nations in 1959. The entire continent became a domain for scientific research rather than a reserve for individual countries or their militaries. Such an agreement was impossible for the Arctic, because unlike Antarctica, much of its territory was already claimed by different nations. While nations and private entities inevitably jockey with one another over Antarctic resources and territory—and try to game the treaty for their own interests—nobody could have predicted the impact of the scientific discoveries that would ultimately emerge.

In this book, polar science will be our lens to see the natural world and the extraordinary ways we have come to know it. We will see how work in the Arctic and Antarctica reveals profound connections among oceans, climate, and all living beings—a delicate network that will shape the coming centuries on our planet. Ice has come and gone for billions of years and, in the process, has sculpted our world and paved the way for the origin of our species. Changes to the poles reveal how topography, climate, and life can change dramatically, often in the blink of an eye in geological time.

Here, the astonishing adaptations of living creatures to extreme conditions reveal how life may have arisen and how it responds to extreme climates. And inside the ice, meteorites that fell millennia ago reveal glimmers of the earliest moments of our solar system.

Ever since humans first set foot on polar landscapes, scientific insights have been gained by researchers often isolated in small field camps or stations, living in environments that can freeze flesh in minutes, and jury-rigging failed equipment to keep life and work going in adverse conditions. Scientific discovery itself has become the product of another kind of science—that of polar exploration. Success comes from employing techniques and ideas developed by Indigenous Peoples who first settled the region thousands of years ago and the European explorers who followed them centuries later. Across time, humanity has expanded its knowledge of the world at the poles through endurance, resilience, serendipity, and tragedy.

Antarctica and the Arctic are warming, and polar treaties are straining as fast as ice melts and species disappear. Our fragile window for understanding the cosmos, the planet, and ourselves is closing. The science of Earth's most remote places and the stories of the people who work there are becoming ever more urgent and relevant.

Once on the ground in 1988, we did not have much time to savor the view. With the changeable polar weather, our first need was to establish shelter. Following the lead of my three companions, Farish, Chuck, and Bill, I started to break out the tents and gear, rummaging through supplies to find ropes, hammers, and tent stakes. Each of the three had a role to play: Farish, the senior professor and former Marine, made the executive decisions; Bill, the experienced field hand, could do everything from cooking a shrimp étouffée in a desert sandstorm to dressing suppurating wounds in the kitchen tent; while Chuck had the keenest eye for seeing fossils amid a field of rocks. My role, beyond bringing a library of Barry Lopez books, volumes on polar history, and field guides to birds, was not obvious.

With tents erected and supplies cinched down, the reality of my situation started to emerge. My relationship to the natural world had been formed 3,000 miles south, surrounded by the greenery of plants and the cycle of dark nights and sunlit days, as well as ready access to other people, food, and medical care. In this new landscape, the sensory cues I once relied on for perceiving time, distance, and the happenings around me were no longer relevant. My fate was now determined by whatever skills I brought, by the three other people with me, and by a long logistical chain to the nearest human beings. With strangeness and insecurity in the air—and my embarrassing lack of camping experience—my tent became a life raft in the immensity of the polar landscape. I found myself perpetually fiddling with the knots, ropes, and stakes in an obsessive quest for the perfect shelter. Somehow, I'd not only have to learn to live in this world but also do the scientific work that had led us there in the first place.

It quickly became clear that science, and in fact life itself, in the Arctic proceeds at the pace set by polar weather, landscapes, and logistics. Mud, boulders, and ice, particularly if there's a headwind, can slow a trek to a few miles a day. Gear that looked great in a catalogue or worked fine in tests in the university quads often fails in subzero temperatures and high winds. Helicopters, planes, and other motorized equipment break in unexpected ways, and spare parts can be half a continent away. Because of these challenges, each week of fieldwork involves several weeks at home planning. And still, wonderfully detailed itineraries, field lists, and objectives go out the window as soon as one sets foot on the ground. As Mike Tyson once famously said about boxing, "Everybody has a plan until they get punched in the face." When plans go awry in polar regions, patience becomes the best strategy—patience with the weather, the people on the team, and, most important, one's own emotional and physical limitations.

During high summer in polar regions, the sun never sets but makes a giant ellipse in the sky throughout the day. Since we arrived more attuned to the light-and-dark rhythms of early summer closer to the equator, a perpetual feeling of jet lag pervaded our first week in camp. With bodies active at night and hungry at odd times of the day, life felt out of sync. There is, however, a nuanced rhythm to polar days. As the light becomes oblique in the middle of the night, temperatures can drop by nearly 20 degrees Fahrenheit. Glacially fed rivers that roared as raging torrents at noon become silent trickles by three a.m. Life, too, recedes and stills. Bees that were buzzing among the plants of the low tundra in midafternoon take to the ground after midnight. Three thousand miles to the south, in places like Chicago, New York, or London, our cues to night and day are found in the levels of light. In polar regions, indicators of time are often sounds.

It's frequently said that we fill the unknown with our familiar experiences, expectations, and fears. In a similar way, the sights and sounds of polar regions initially evoke sensations from home. Melting water that rushes through glacial rivers produces a chorus of shrieks or high-decibel roars that recall sirens, elevated trains, or the din of Chicago's Loop. I am often startled by sounds that evoke sizzling bacon on a skillet when I hear glaciers move. Booms that sound like thunder make me reflexively look up to the sky until I remember that they actually emanate from below as ice ruptures. Rubble that was moved by flowing glacial water appears, at first, like a bulldozed construction site. The combination of snow, wind, and light has an almost supernatural effect on ice, etching beveled edges and sharp cornices to look as if they were carved by modernist sculptors. We conjure an informal cartography for the terrain that refers to familiar places: mesas, peaks, and ridges are named after campus buildings, bell towers, or caricatures of features of people, animals, or famous places.

After two weeks on-site, the polar world comes into clearer focus, and what was initially disorienting becomes liberating. There is something almost magical about living in an environment where the sun never sets for a month or more while being disconnected from the rest of humanity. Running streams exiting melting glaciers hold water so pure we drink it unfiltered from the source. Every babbling glacial brook could be a water fountain or, for extraordinarily hearty souls, a bath. The continuous light of the sun, coupled with the vastness of the landscape, brings an almost boundless metabolic energy that powers long hikes, intense conversations, and deep companionship. The world becomes small and intense—defined by the colleagues on the expedition, the supplies and gear that we bring, and the rocks, rivers, and tundra that are accessible on foot. With these local and intimate interactions, the landscape—its shapes, textures, and rhythms—gains a personality, almost like a member of the expedition. The shifting weather and light of the Arctic and Antarctic shapes our work, lives, even our emotional states. With no news, internet, or external sources of information, the main distractions in camp are the ones we carry inside of us.

Much of what makes polar regions special is their relationship with the sun. Documentary filmmakers have had a field day with some of the misconceptions about this relationship, which most of us learn in elementary school only to forget by middle age. One film crew went to the commencement at Harvard University and asked students and faculty the question: What causes the changes of the seasons? Virtually everyone said that Earth is closer to the sun in summer, leading to its relative warmth, and farther away in winter, making it colder. Of course, they could not have been more wrong.

Earth's axis is tilted, and as the planet orbits the sun, light hits the surface at different angles at different times of the year. During winter, a region receives oblique light; in summer, the light hits more directly. Sunlight that travels at a low angle to Earth loses energy by the time it gets to the surface. Hence it is colder during these winter times. The opposite is true for summer: light that hits the surface closer to a 90-degree angle arrives with more energy than oblique light, making the environment warmer.

The same relationship between the angle of the sun's rays and energy explains the differences between the poles and the equator. The sun is, on average, about 93 million miles away from Earth. That vast distance means that the equator and Earth's poles are, for all intents and purposes, equidistant to the sun. The distinction between the poles and the equator lies in the angle at which the sun's rays reach the surface: light hits polar regions at a lower angle than at the equator. Accordingly, sunlight that arrives at the Antarctic continent at an angle of 30 degrees on a cloudless summer day carries half as much energy as that which falls on the surface near the equator. Because polar regions receive far less energy from the sun throughout the year than other parts of the Earth, the poles are low-energy places. Creatures that live there either find new ways to produce energy, conserve it, or learn to live with little of it.

Because the Earth's axis is tilted, the poles are continuously dark at the height of winter and light in summer. One definition of the Arctic and Antarctic Circles is the latitude at which the sun never rises, or sets, at least one day a year. The energy on the surface of the planet ultimately comes from the sun, and polar regions have no incoming solar energy for months at a time. Living things run on a kind of fuel reserve during these dark periods.

The reaction of the polar environment to light is an exemplar of its susceptibility to change. There's an illuminating analogy from chemistry and physics. Chemists talk about a system being "metastable" when any small change can push it to a new state. Imagine a ball delicately balanced on a stick—any tiny nudge can send it moving in a different direction. In an analogous way, small shifts in climate can transform polar regions. Because these regions function with so little energy from the sun, relatively subtle alterations in climate can move polar regions from an icy state to one lacking ice altogether.

Polar regions encompass 8 percent of the total surface of the Earth but hold an outsized influence on the state of the planet. Almost 70 percent of all the planet's fresh water is frozen in ice. On land, permafrost in the polar regions holds 1,600 billion tons of carbon—roughly double that in the entire atmosphere today. Locked in the soils and ice of the poles are clues to our past and things that will shape our planetary future. Every milestone of human evolution, from the origin of our species to the establishment of our social structures and technologies, arose during a time of ice at the poles. Freezing and thawing over millennia, the Arctic and Antarctica are like vaults that hold our planet's heirlooms. When polar regions melt, the vaults are thrown open—ancient water, carbon, and microbial life return to the surface to shape and change the world.

Layers of history permeate these landscapes. Valley walls hold layers of rock that reveal billions of years of changes on the planet. The landforms of hills, valleys, and mounds of gravel speak of more recent ways ice and wind have sculpted the surface. Ice, light, and wind are like an invisible hand that has carved these worlds over thousands of years. The ice itself contains atoms that tell how it grew and shrank over millions of years, while bubbles and ash caught inside the ice are relics of ancient worlds. And scattered across the surface of the tundra and coastlines are artifacts of human history, from the dwellings of the Indigenous Peoples who first settled here to implements left by European expeditions hundreds of years ago.

Other histories lie in this terrain. In the three decades since that 1988 trip to Greenland, Farish, Chuck, and Bill have each passed away, and much of the snow and glacial ice we trudged through in those days is now lost to time or climate change. Back then I was a student learning to find fossils, to figure out how to live in the wild, and to discover that success needs to be measured in years rather than weeks. Veterans of polar work—pilots, research personnel, or support staff—speak of the ways the landscape becomes part of their blood, like the whalers in Herman Melville's *Moby-Dick*. Now having led my own expeditions to polar regions for several decades, the work has become a part of me too. And not only figuratively. As the saying goes, every wrinkle tells a story. Aching joints, scars, and weathered skin are records of encounters with ice, rock, and river crossings and tumbles too numerous to count. But the converse is also true: our lives become part of the landscape itself.

I departed on our 2002 expedition to Canada's Ellesmere Island a week after my infant son arrived for adoption. Our team was in the middle of what became a six-year hunt for fossil evidence of the earliest fish to walk on land and was at a critical phase. It was also one of the wettest field seasons in our experience—we sloshed through the mud around camp and geological outcrops for the entire time. The team returned home with only soiled gear and bruises that year. Six years later, after finding that fossil fish with arms (known as *Tiktaalik roseae*), we returned to the 2002 site for another look in better weather. During one traverse, I happened upon my old footsteps locked in the dried mud. With my son now in first grade, I stood before those six-year-old prints made by the earlier me, the one who walked here cold, wet, and questioning his judgment. By capturing that fleeting moment in time, the prints embody the frailty of the place. Slow-growing tundra mosses, lichen, and wildflowers can be damaged by footsteps captured in dried mud for decades.

These fragile landscapes reveal our own vulnerabilities. Rapidly changing polar regions will increasingly affect lives across the entire globe. Seeing our connection to these places, and appreciating what we have learned from them, means coming to grips with the surprisingly complex material that makes the Arctic and Antarctica so special and important.

# CHAPTER 1

# ICE IS HOT

Strapped to the netting inside an Air Force LC-130—a whale-shaped Cold War–era propeller plane that is older than virtually all of its passengers—it is hard to escape the fact that working in Antarctica is 180 degrees different from working in the Arctic, both literally and figuratively. In 2018, our eleven-hour flight to McMurdo Station began in Christchurch, New Zealand, where we had spent two days receiving extreme-weather clothing and getting introduced to the U.S. Antarctic Program. I was one of thirty passengers, all of whom were wearing the program's iconic parka, known affectionately as Big Red. Big Red is a warm home away from home in the field, a giant blanket of a coat with enough pockets to lose small items for an entire field season. Now, Big Red was our survival gear in case the plane crash-landed on ice during the flight over the rough seas of the Southern Ocean.

Donning earplugs to deaden the propeller noise, sitting at a distance from my neighbors because of Big Red's girth, it was nearly impossible to hold a conversation. I gleaned from shouts that my colleagues on the flight included ice drillers, fire-safety personnel, and research scientists. Working in the Arctic, something that felt so extreme in the early days, now seemed like a quaint family camping trip. Antarctic fossil hunting would be more akin to a military operation.

As we deplaned the LC-130, the hatch felt like a portal to a new existence: from the dark confines of the belly of the plane we emerged into a cold, dry, and searingly bright world. Southern polar rookies like me savored our first lungfuls of Antarctic air and wandered around the ice landing strip relishing new vistas while the staff tried to herd us into the transport to the base. Slipping on the hard-packed ice, I soon realized I would need to learn new ways of walking, living, and working in this environment. I'd spent my career thus far primed to look at rocks for fossils. But in Antarctica, there was a new factor to consider in the field. The southern polar region rises thousands of feet above sea level. Antarctica is a continent and does not have relatively warm ocean waters penetrating its interior the way much of the Arctic does. Consequently, Antarctica is colder, windier, and drier—and most important, icier—than the Arctic. The U.S. Antarctic Program calls the time in Antarctica being "on the ice" when on base or in the field. Whether preparing at home, training at the station, or sleeping in a tent in the field, Antarctic work is defined by a relationship with ice.

Humanity's experience with glacial ice, both on mountaintops and at the poles, is a story of adventure, tragedy, and scientific observation. For most of us, familiarity with ice derives from encountering patches or pieces of it—skating rinks in winter and cubes in summer drinks. These experiences can hardly prepare you for the workings of ice the size of small towns or even entire continents.

## The Beauty of Water

Being on an American expedition in Antarctica often means spending time at McMurdo Station, the largest scientific outpost on the continent. Mactown, as the locals call it, sits on an active volcano, while its airstrip lies atop a shelf of ice that borders the base. Known as the Ross Ice Shelf, it is as large as the Mediterranean Sea and is nestled between mountain ranges on one side and smoking volcanoes on the other. The rocks around McMurdo are dark basalts, cooled magma bombs that were ejected from the volcanoes in years past. The location of the airstrip has had to creep farther from the station in recent years as climate warming thins the ice below it.

Since the ratification of the Antarctic Treaty in 1959, seventy scientific bases have been constructed in Antarctica by twenty-nine different countries. The total population of the continent never exceeds 4,000 people. McMurdo is one of three bases run by the U.S. Antarctic Program. It houses over 1,000 people in the height of summer and about 300 in winter. The South Pole Station is 900 miles away and sits at an altitude of 10,000 feet. A big summer event for scientists and staff at the pole has been the Round the World Run—nerd humor for a kilometer-long costumed jog or ride around the globe done in subzero temperatures. Each country's base becomes its own world with a distinctive culture. The food options on French and Italian bases—from pâté to risotto—are as different from McMurdo's meatloaf, pizza, and nachos as are the field gear and extreme-weather clothing donned by researchers. American researchers in Big Red look like five-year-olds on their first sledding trip when compared to the New Zealanders in their slick black-and-green parkas.

McMurdo is the port of entry for most expeditions on the ice, and it feels like a waystation similar to the fictional Mos Eisley spaceport in *Star Wars* or Rick's Café from *Casablanca*. The entire

Round the World Run, including those less inclined to run, in 2010

population is temporary: those in transit to field sites, such as me and my team, interact with people who are on base for six months or more and maintain and support its operations. Many people are waiting for flights into the field, others for flights home. Each year, several expeditions get stuck on the base for months, or never get into the field because of equipment failures or weather changes. Some arrive at McMurdo only to find their entire scientific project scuttled because a part for an experiment didn't arrive or the weather where they want to work is too forbidding.

Ice was my silent partner on base—I saw it from the panoramic windows of the science library. I'd trained in the base's hangars to camp and live on the ice, but I wasn't allowed to actually be on it until all our gear was checked and we were certified to go into the deep field. I spent weeks walking around McMurdo with a clipboard, taking notes on topics ranging from snowmobile maintenance to environmental safety, while I also checked and rechecked field lists of equipment, food, and supplies.

Bad weather had backed up flights into the field, and the population of McMurdo swelled to the point where finding a seat in the cafeteria at dinner became a minor expedition itself. One evening I found an empty spot at a table with a mix of glaciologists and ice drillers engaged in an animated conversation. One of the scientists was ruminating on how the delayed work on the West Antarctic Ice Sheet was going to reshuffle their plans. His easy demeanor was as captivating as his knowledge of glaciers, so I hung around to chat after we finished our meal.

Sridhar Anandakrishnan trained as an electrical engineer in college in preparation for a career designing equipment for scientific research. He settled at the University of Wisconsin to pursue his engineering degree, but lacking any financial support, he was stuck looking for a job to get him through the summer months. A roommate scanning the newspaper had found an ad for work in the geology department designing and building a portable seismometer to measure movements of the ground's surface in remote field locations. Lacking any other opportunities, Sridhar took the summer gig. After he built the gadget, the lab's leader came to him with a proposition: Could he join his team for an expedition on the ice in Antarctica? The group wanted to use Sridhar's seismometer, but because he knew the details of its construction intimately, he was the only one who could repair it if something went wrong.

Sridhar describes the moment he stepped foot on Antarctica as "the end of the story." He was hooked on polar research and was set on a new path in life. Shifting fields, from engineering to ice, Sridhar needed to take a crash course in geology. But it all paid off when he found himself going into the field and learning about the ice firsthand. Working on the ice was an intellectual and visceral thrill for Sridhar. He gleams as he says, "Where else can you learn about our world while living in a camp eight hundred miles from the nearest humans?"

By the time I met him at dinner, Sridhar had become a professor at Penn State University and had worked for more than thirty seasons on glaciers in Antarctica and Greenland. He specializes in using high technology to probe into the ice and explore its interaction with the ocean waters and bedrock below. As I got to know him, I became more comfortable asking naïve questions. Seeing my ignorance about the complexity of glacial ice, Sridhar smiled and casually observed that I needed to understand that "ice is hot."

Sridhar's glaciological Zen koan explains what is so weird and important about ice. The physics of water molecules underlies much of our world. Water is an unusually important substance because it exists as a solid, liquid, and gas within a relatively narrow range of temperatures. Obviously ice near water's freezing point of 32 degrees Fahrenheit is not hot by any measure of temperature. But because of the physics of the molecule, ice is hot in terms of the temperature required to melt it into a liquid. A steel bar melts

Sridhar Anandakrishnan in his element

at a temperature of roughly 2,300 degrees Fahrenheit. Steel needs to be heated to extreme temperatures to go from a solid to a liquid, let alone a gas. Meanwhile, the melting point of ice, at 32 degrees Fahrenheit, means that in natural settings it is much closer to the transition from its solid to its liquid state than a material like steel. Only a slight change in temperature is needed to shift the water from a solid to a liquid. And because pressure and temperature are related physically, ice under pressure melts at an even lower temperature. Ice lining the bottom of a glacier sits at such high pressure that it is right on the cusp of melting.

The fact that "ice is hot" spawned one of the most vicious scientific debates of all time. Far from being an esoteric academic feud, this conflict illuminates the surprising things ice can do.

Louis Rendu, Bishop of Annecy in France, developed interests in both the creator and the glacial ice in the mountains near his home at the border of Switzerland. In 1840, after cogitating about glacial movement and looking at the shapes of the glaciers around him, Rendu supposed that glaciers moved like a lump of honey on an incline. As a viscous substance, honey stretches and deforms as it descends, spreads, and follows the general topography of the surface over which it flows. Using this analogy, Rendu believed that glacial ice fills valleys and all the nooks and crannies inside as it spreads. Bishop Rendu published his idea in a treatise titled *Theorie des Glaciers de la Savoies*, and noted that glaciers have "some sort of extensibility" and "may move like soft dough." In 1841, the physicist James Forbes was told of Rendu's theory and essentially popularized it by announcing his own "viscous or plastic theory of the movement of glaciers."

Entering the conversation next was Lord Kelvin (born William Thomson), who later became famous for his discoveries in physics, including thermodynamics. He worked with his brother on theories of glaciers. They noted that at high pressures ice can melt at lower temperatures than normal. This meant that the ice at the depths of a glacier, which sits at high pressure, could melt into liquid water despite being cold. The liquid at the base could serve as a lubricant between the ice above and rock below, enabling the entire glacier to glide over the bedrock. Melting will happen anywhere in the glacier where ice sits at high pressure. The interior itself can melt and break apart, and the different parts of the glacier move downhill at different rates. The brothers envisioned that glaciers don't move as a single block of ice; with different parts of its interior melting as it sits under pressure, ice could crack and flow downhill, and different pieces could fuse as they came back into contact and refroze.

The physicist John Tyndall, a fiery Scot, was as attracted to Lord Kelvin's approach as he was utterly appalled by that of Rendu and Forbes. It didn't hurt that Tyndall also held a deep and long-lasting antipathy to Forbes. By the time Tyndall became interested in glaciers, he had already attacked Forbes's theories on other topics—even one on, of all things, musical instruments. Tyndall never lost an opportunity to savage Forbes and make him look like an incompetent fool. Turning to glaciers, Tyndall thought Forbes's use of the term "viscous" wildly off base. To him it was obvious that glaciers didn't move as a viscous, honey-like substance—they cracked, slid over melting waters at their base, and fused again anew as Lord Kelvin suggested.

In retrospect, Rendu, Forbes, Kelvin, and Tyndall were each correct to some degree. Glacial ice, it turns out, can do it all. It is not one thing—it is many things. Part solid, part liquid, and part plastic, glacial ice can slide, bend, ooze, melt, and refreeze. Whenever there is liquid water, strange things can happen. With ice at pressure becoming liquid water, parts of a glacier can move relative to other parts like a thick gel. Glaciers can flow around obstructions by splitting like a river around a boulder, or deep layers can be pushed to the surface.

Polar landscapes have captivated humans for millennia with their varieties of ice. It's been said famously that Inuit have around fifty-two words for ice and snow. While the specific number of words in that claim is controversial, the Inuktitut language uses varied combinations of words and phrases to describe ice's shape, form, strength, and usefulness for traveling, cooking, or hunting. Polar ice can take almost infinite shapes as it crystallizes, moves, and melts. The area around McMurdo Station exhibits this diverse world of ice in microcosm. The mountain ranges near McMurdo look like a dessert that would have pleased Rendu; the ice covers the mountains like a pure white syrup, filling canyons and valleys as it flows. Elsewhere ice looks like glass that has shattered into shards, strips, or chunks ranging in size from a small car to an entire skyscraper. The nearby sea is covered by ice that looks like a layer of foam at one time of year, a quilt of polygons at another, and a folded blanket at still another. Part of the awe of seeing a polar vista comes from imagining how a chemical formula as simple as $H_2O$ can underlie worlds of shape and movement that are almost magical in their variety.

Ice comes in structures as different as individual crystals and sheets the size of North America. Hidden within this complexity are secrets to the ways it sculpts landscapes and changes our entire world.

## Living on Ice

I was drawn to polar regions during a hunt for the most ancient fish to have walked on land. While it may seem strange to look for fossil fish near the poles, the movement of the continents during Earth's history means that rocks that formed near the equator millions of years ago can now be found at high northern and southern latitudes. After my colleagues and I discovered the fishapod—a

375-million-year-old fish with arms and legs—in rocks in the Canadian Arctic in 2004, we later found scientific papers attesting to similar rocks in Antarctica. Antarctica is divided by the Transantarctic Mountains, which, as the name implies, extend across the entire continent. Their summits are the rocky equivalent of "the tip of the iceberg." The mountains rise over 14,000 feet from their base, but because the ice sheet is miles thick, they extend only a few thousand feet above the ice surface. In 1914, Robert Falcon Scott's team discovered fossil fish scales in some of these rocks. In the 1960s, New Zealand teams led by Sir Edmund Hillary and, two decades later, Margaret Bradshaw, found more fish bones and teeth. In more than 150 years of Antarctic exploration, Bradshaw was a pioneer—one of the few women to lead deep-field teams. In 2018, my team and I were off to follow in Bradshaw's footsteps to explore these mountains for fishapods. But as with every expedition on the continent, the real player is the ice and the way it influences our world.

Each deep-field team has a mountaineer assigned to it, somebody who can train the team on ice tactics before the trip and help keep the expedition safe in the field. While I was the scientific head of the trip, our mountaineer defined what was doable—a conversation he and I would have over coffee in the tent each morning while the rest of the crew slept. I met Forrest McCarthy on the advice of longtime Antarctic veterans while I was preparing for the expedition at home. He has startling blue eyes that stare right though you and an almost childlike enthusiasm for work on the ice. By the time we met, Forrest had guided dozens of expeditions and had worked for the search-and-rescue teams in McMurdo and at Denali in Alaska. Beginning in 2017, a year before this particular trip, Forrest and I went over our gear needs and sat with aerial photos of the sites to assess how to get in, get the work done, and get out safely. Our plan was to fly in with our supplies, fuel, and snowmobiles to the middle of a flat glacier known as Deception Glacier, then use the vehicles to set up camp at the base of the mountain that held promising rocks for fossils.

After two weeks of a mind-numbing routine training and shuffling field gear at McMurdo, Forrest came up to my table at breakfast with a look in his eyes that told me our drudgery was about to end. He'd just learned that we'd be heading out to do the fabled "field gear shakedown and mock crevasse rescue" the following day. I knew what that meant. One of our first experiences with Antarctic glaciers wouldn't be on the ice—it would be *inside* it.

We filled the snowmobiles with fuel, packed our camping gear, and departed McMurdo the next morning. After setting up camp on a flat and stable patch of ice, making ice shelters, testing the tents, and firing up the stoves, we set off to a crevasse. The mountaineers use a small crevasse near the training site for mountaineering lessons—a playground version of the treacherous crevasses around base. Even so, looking down a 20-foot crack in the ice is vertiginous. Crevasses are a major cause of fatalities in the region; these cracks can descend 200 feet and be invisible from the surface. Two years before our trip, Forrest had extracted the remains of a famed researcher who died when he fell into a hidden crevasse. He had been one of the world's leading experts on glacial crevasses and was snowmobiling across the ice near McMurdo. The student driving behind described him as having simply disappeared from the surface. Crevasses can be deceptive because cracks at the surface can be covered by thin, ephemeral snow bridges, which can collapse under any small weight. That tale was on my mind as Forrest brought us to the crevasse to train. The plan was to lower a member of the crew into the ice while the team on the surface learned to build rigs to pull the person up.

Forrest walked us through the sequence of knots, pulleys, and carabiners needed to haul a large person up to the surface from deep in the crevasse. The pulley system is an amazingly simple and

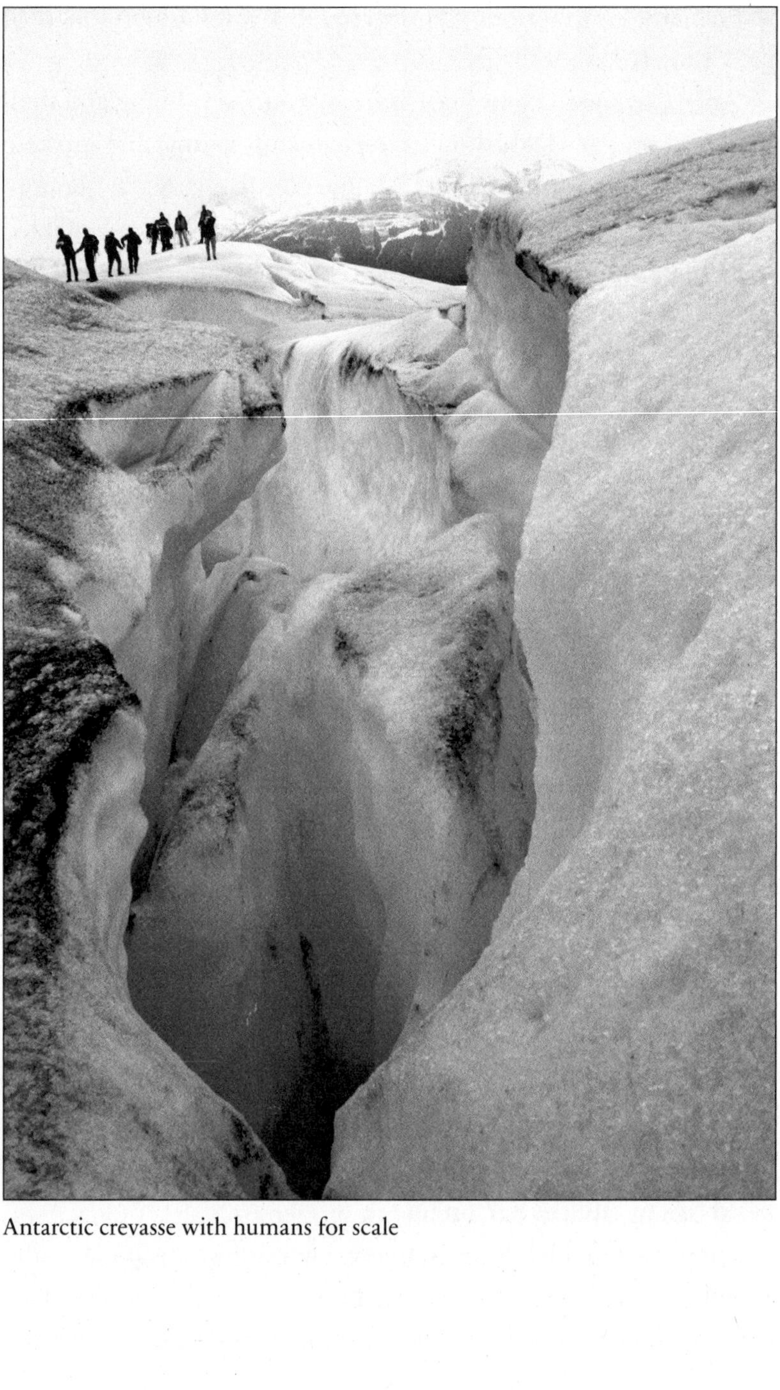

Antarctic crevasse with humans for scale

lightweight contraption—the mechanical advantage to extract a fully grown adult who has fallen 100 feet below the ice surface can be assembled from items kept in a small daypack. But a misplaced loop, a hitch tied too loosely, or a frayed rope can mean the difference between life and death. With that fact in mind, I memorized the sequence of knots and pulleys until I could build the system with my eyes closed.

Tim, one of the burly Australians on our crew, was a spelunker in his spare time and did not have a hint of the claustrophobia that the rest of us felt at the idea of squeezing into a crack in the ice not much wider than Big Red. After Tim gleefully volunteered to be lowered into the ice, we attached ropes and carabiners to a harness clasped around his waist and thighs. Once he got the OK from Forrest, he descended from view looking like a puffy red caterpillar undulating into the earth.

Cries of awe, punctuated by occasional irreverent Aussie slang, emanated out of the crevasse. Tim's shouts from 20 feet down were as if he were undergoing religious ecstasy. Once we hauled him up safely, the team needed to hear what the fuss was all about.

As he described his experience, Tim's eyes glowed like Charlton Heston's as Moses holding the Ten Commandments. The trip inside ice was like crossing a gateway to an exotic realm of new hues, shapes, and contours. Tim's world became awash in a sea of blues of different shades that got darker with each passing foot of depth. Frozen water absorbs all other wavelengths of light, leaving blue reflected. Light flowing through the ice illuminates ice crystals in the walls that take a cornucopia of shapes, from slender rods to crystals the size of the palm of your hand. With walls of ice forming pillars, folds, and arches, each with numerous layers, the place felt as if it were designed by an architect. Sounds changed as well. The rustling of the wind and human voices are replaced by sounds of your own beating heart and clipped breath, and the crinkling of ice that resembles gentle wind chimes. The utter lack of evidence of life, the narrow and multiform spaces wrought by nature, and the filtering of the light from above made the interior of the ice feel sacred. It was like a blue chancel of an icy Gothic cathedral, a space hallowed by its isolation from the populated world nearby.

The interior of the ice reveals the ways glaciers change with depth. Falling snow each year covers the packed snow that remained from the previous one. Air separates the flakes, which are so dry they move like a mound of feathers in a breeze. If the weather stays cool, the ice builds year to year, with the fluffiest snow on top and ever denser ice and snow in deeper layers. With even more accumulation, the snow below gets squeezed into ice at higher pressures. Because the melting point of water is related to both temperature and pressure, the ice at depth sits close to its melting point. Any small change can cause these layers to melt and later refreeze. Over time, this ice compacts into a rock-solid substance with only tiny air bubbles inside.

When the ultra-compact ice from the depths is brought to the surface, it looks like a shiny version of an aquamarine. This ice, known as blue ice, can be a torment for polar researchers. It looks like a gentle azure pool of tropical water from a distance. But looks can be deceiving. Blue ice is so frictionless that even the simplest motion becomes fraught. There is absolutely no way of getting enough traction to move on a flat patch of it. Crampons often don't work because they cannot etch into the ice's rock-hard surface. Snowmobiles caught on a slope of blue ice have been known to slide hundreds of feet downhill.

Blue ice tells the story of movement—it forms at depth in a glacier and is heaved to the surface in certain locations. Like glaciers everywhere, ice sheets in Antarctica are always flowing from areas of high elevation to lower ones. Occasionally, an ice floe will hit an obstruction such as a mountain range and blue ice will extrude upward from the deep layers of the glacier. This ice can rise vertically hundreds, if not thousands, of feet. The deepest layers hold the most dense, hard, and ancient blue ice—some of it is more than 6 million years old. While it is one of the greatest hazards to work on the surface, blue ice in its antiquity brings sublime joys. Melting this ice for drinking means consuming water from snow that fell when Neanderthals roamed the Earth.

We camp on ice, melt it for water, make accommodations with it as we work, eat, and sleep. Ice becomes part of the fabric of our lives like the air we breathe and the food we eat. Ice is our highway to the rocks on mountains that poke through it. Ice also influences the weather and the entire landscape around us.

At the base of a mountain in Antarctica there's the pervasive threat of disastrous winds, sometimes over 100 miles per hour, that not only can stop work but can wipe out a camp. Ice at elevation makes the surrounding air cold and very dense. That dense air can flow downhill and, gaining speed as it descends, generating intense windstorms in the floor of a valley. Forrest was famously caught in an Antarctic gale that sent his team's snowmobiles flying though the sky. Of course, given Forrest's experience, the machines on our trip were parked far away from tents.

Extreme winds are like an invisible sculptor of the world at our feet and in the distance. Fields of rocks, scoured by blowing air, are scattered across Antarctic valleys. The direction of the prevailing winds is etched on the rock surfaces. Over centuries, this erosion carves a range of shapes in the rocks and boulders, from Corvette-like wedges to sponge-like, hollowed-out husks. The valley's features also show the action of centuries of continuous gales. Cliff surfaces bear scalloped edges, hollows in rock walls, and pillars of sandstone—all features crafted by wind.

Wind and ice together shape entire landscapes and create new ecosystems. A set of parallel valleys, covering almost 2,000 square

Antarctic Dry Valley with a windblown rock in the foreground

miles, lies 40 miles west of McMurdo Station. While these valleys are flanked by ice, their walls and floors are pure bedrock with barely a patch of ice inside. The area looks like the surface of a rocky, lifeless desert. In winter, temperatures here can drop to –90 degrees Fahrenheit. The region is so cold and barren, it is used to model the geology and possibilities for life on Mars. The landscape of these valleys, known as the Dry Valleys, appears as a paradox. If ice lies on the mountains above the valleys, and their interior gets extremely cold, why is there no ice within the valleys themselves?

Small mountains rise over 8,000 feet along the edges of the valleys. These ridges serve as natural dams that block the movement of glaciers into the valleys below. Instead of entering the valleys, the ice remains on the tops of the ridges. There, cold air accumulates above the ice. When the conditions are right, this cold and dense air rips into the valley at the speed of a major hurricane. These winds carve the valley walls and keep any snow from accumulating in areas of low relief. The winds make for a cold desert oasis on the vast, ice-covered continent. Picture the Grand Canyon surrounded by a world of ice. The mythological names of the mountain ranges that line the valleys—Asgard Range and Olympic Range—conjure the awe they inspire.

The winds throughout Antarctica carve the rocky landscape and sculpt and resculpt the ice surface. The surfaces of glaciers are covered with hummocks of snow that can rise as high as six feet. Known as "sastrugi," these features loom large in polar expeditions. Explorers who tried to sledge to the North and South Poles recount tales of being slowed by having to push their sledges over these humps. Blowing snow, the formation of sastrugi, and the regular movement of ice make the surface of a glacier feel like an ocean with moving waves and surfaces that change from day to day.

If there's a Murphy's Law of paleontology, it is that the best fossils are often found at the tops of mountains. Our fossil site was no exception. With moving sastrugi, routes up from Deception Glacier to the mountain changed almost daily. Having learned to walk, live, and climb on polar ice, we'd ascend through millions of years of geological time as we moved up layer after layer of rock with our eyes close to the surface to find fossils. Our reward was the fossil bones and teeth of 380-million-year-old sharks, armored fish, and evolutionary cousins of the fishapod we found in the Arctic. An ancient tropical river once teeming with life is now a rock layer lying atop an Antarctic peak surrounded by ice that extends hundreds of miles to the South Pole.

Polar landscapes expose the transience and precariousness of our own existence as well as that of the entire living world. Change is everywhere on the ice. Seeing the impact of this dynamism on a planetary scale means developing a new set of eyes.

The tops of these mountains contain rocks formed in ancient rivers and streams and hold fossils of ancient fish and sharks.

## A Glacial Pace?

Eric Rignot's path to a career on the ice was not through mountaineering or geology but through mathematics. Math, as he was told by his teachers in France, was a ticket to interesting careers. And in his case, it was—the discipline opened doors for him to explore his passions in engineering, aeronautics, and astronomy. By the mid-1980s, Rignot became fascinated by the newly burgeoning field of satellite imaging. Landsat missions had been under way since 1972, imaging the Earth with ever greater precision with each new satellite launched. Stimulated by the possibilities of this new science, Rignot gravitated to the Jet Propulsion Laboratory (JPL) in Pasadena, California. Famed for its satellites and space probes, the JPL was also home to some of the world's most advanced remote-sensing work.

When Rignot joined the team at JPL, his boss told him to work on ice. Rignot wasn't sure. As he told me, "I actually thought my boss didn't like me and that's why he pushed me to work on glacial ice." But soon, Rignot's time at JPL convinced him that satellite-based remote-sensing techniques could revolutionize our understanding of the ice sheets in Greenland and Antarctica. Ideally, satellites could allow scientists to see and measure an entire glacier at once, giving access that might otherwise be impossible because of the difficult logistics of polar work. Also, with the technology that was emerging at the time, satellite imaging could allow him to see ice in new ways, even under cloud cover.

Although he started his work with skepticism, it turned out he was in the right place at the right time, doing the right thing. In looking back at the start of his career three decades later, Rignot said, "Pretty soon, I realized that I stepped into a gold mine of information from remote sensing techniques that were changing completely the ways we were looking at ice." Much of what we had thought we knew about glaciers and their response to changes in Earth's climate turned out to be wrong.

In the early 1990s, astronomers at JPL devised a satellite technique to measure subtle movements of the Earth's surface. Radar pulses sent from a satellite in space to the planet below are reflected back to the satellite. The timing of the round trip of a pulse gives a measurement of distance between the surface and the satellite detector. If the same spot is measured several times, the signals can be combined to reveal the velocity at which objects below are moving. The technique can be so precise that changes on the order of a millimeter can be measured from space. To Rignot, this technique was like a new pair of glasses—a whole world of ice, and its movement, could be explored.

Rignot and his team at JPL, NASA, and international space agencies scanned the satellite data to make detailed measurements of the movement of ice across the entire continent of Antarctica between 2007 and 2009. In terms of scale, this effort is akin to looking at the water flowing across the entire continental United States every day over a two-year period. The imaging generated an

Eric Rignot

enormous amount of data, which was then crunched and visualized with powerful computers at the University of California, Irvine. When the satellite data were rendered in graphical form, they revealed the movement of ice across the entire continent. The glaciers obviously aren't static, but the ways they move are surprising. Streams of ice, which from space look like rivers some 5 to 10 miles wide, flow from the center of the continent thousands of miles to the coast. As these rivers of ice flow, they break up to form tributaries of ice that stream toward the sea. These corridors of moving ice can be 500 feet thick and travel over half a mile in a single year.

Just as a map of North America from space shows rivers of water flowing from the Continental Divide to the oceans on both coasts, a satellite image of the ice of Antarctica reveals rivers of ice flowing from the continent's interior to the ocean around it. And this flow has been happening for millennia.

The satellites reveal that the entire ice sheet of Antarctica is continually on the move from the center of the continent to the ocean. If you put a boulder that could be traced year after year at the South Pole, it would end up in the ocean within one hundred years. When the ice hits a mountain or some other obstruction, the conveyor belt continues and it flows around the obstacle while some ice from the deepest layers wells up. These regions also develop cracks, crevasses, and ice cliffs where the ice folds and bends.

Satellite discoveries led Rignot and his colleagues to a new question: If ice sheets are always flowing toward the coast, often rapidly, what controls the speed at which they slide, grow, or melt?

Glaciologists have envisioned the size of a glacier to be set by the balance of processes that add ice each year and those that remove it. Moreover, ice will flow like a river via the force of gravity from higher elevations to lower ones: new snow at higher elevations compacts year over year and flows downhill. As the glacier flows downhill, ice is lost to melting, crumbling at its margins, and the direct evaporation of solid ice in the wind. Warmer conditions mean less snow accumulation at the top, and more melting, cracking, and evaporation at the bottom. All else being equal, the glacier's movement downhill will speed up as it shrinks. The converse is true in cooler conditions.

New snow that persists each year adds volume to the glacier, while melting throughout its surface and downstream edge reduces it. Taking this notion to its logical conclusion, the way glaciers should shrink in a warming climate is that ice should melt at the surface and the water should flow into the ocean. When the planet cools, the melting and water flow should stop while snow is added to the surface, allowing the ice to grow.

But satellites were revealing something different. Pings of lasers sent from space to the ice don't only provide a snapshot of the condition of the ice, they reveal how the height of glaciers changes over time. Taking these measurements over months and years can show where and how fast the ice is shrinking and melting. These refined measurements reveal that new snow clearly is added each year. But most of the changes in mass of the ice sheet, and the glaciers that extend from it, do not happen by melting on the surface. Looking at the ice change suggests only 5 percent of ice loss in Antarctica happens this way. If melting on surfaces the size of a continent does not contribute significantly to changes in the size of the glacier, what does?

The laser signals were revealing that the loss of ice in the glaciers wasn't spread across the continent; it was highly localized. The greatest change in the height of the ice, meaning the fastest melt, happens along the coastline, where the ice meets the ocean. Changes at the coast then propagate inland to the center of the ice sheet. The "business end" of the ice sheet is where it meets the warm water of the ocean.

Still other missions added new pieces to the puzzle. The Gravity Recovery and Climate Experiment (GRACE) missions, a collaboration between NASA and the German space agency, DLR, were initially launched in 2002 with a clever idea. Two satellites (nicknamed Tom and Jerry) were placed in orbit 310 miles above Earth. They circle the globe 140 miles from one another, one trailing the other (presumably Tom is chasing Jerry). Each satellite shoots a beam of microwaves to the other, enabling the researchers to measure how far apart they are. The resolution that these microwave measurements provide is nothing short of astounding—they can detect changes in distance of a micrometer, which is one thousandth of a meter. As the satellites orbit the planet, they encounter changes to the Earth's gravitational field. Areas with lots of mass, such as thick sheets of ice, exert a greater gravitational pull on the satellites than do areas with thinning ice. As the satellites move over areas with thicker ice, the one ahead will be pulled away from the one behind.

Then, as the lead satellite passes this area, it will slow down again. Now it's the trailing satellite's turn to be sped up by the gravitational pull, only to slow down as it passes the area with high mass. While these two satellites accordion away from and nearer to each other by the pull of gravity, the microwave beams are recording the distance between the satellites. The changes in distance allow researchers on the ground to determine where there are areas of high gravity (hence, greater ice mass) and lower gravity. These satellites monitor gravity and mass 24 hours a day, 7 days a week, 365 days a year. We now have a continuous record of changes in the mass of polar ice over time. The GRACE missions send data that Rignot calls a "Weather Report" of the changes to the mass of ice in Antarctica and Greenland.

With satellites able to measure the mass of the ice in different places, the speed of the ice's movement, and places where the ice is thinning, the dynamism of the ice sheets is coming into clearer focus. At the coast, ice meets the sea and liquid water interacts with solid ice. The melting here is the most dramatic because the warmth of the ocean causes the edge of the glacier to thin, collapse, and crumble into the sea. Melt can also happen from below when the relatively warm water of the ocean flows underneath the glacier between the bedrock floor and the ice. This melt can set off a chain reaction: the more the leading edge collapses, the faster the ice streams will bring ice from the interior to the coast, leading to more ice interacting with the warm ocean waters. Some coastal glaciers, such as the Cadman Glacier, have lost as much as 60 feet of height—the equivalent of a five-story building—in a single year.

Scientific history offers perspective to these recent satellite discoveries. In the 1800s, mountaineers and natural scientists noticed strange patterns in the ice and rock of alpine landscapes. Seeing that the valleys that hold glaciers are larger than the ice itself, they surmised that at some point in the past the glaciers had filled the valley and later retreated. The alpinists also were intrigued by massive rocks strewn across the terrain that were unlike any others in the area. With these facts, they hypothesized that over long periods of time, glaciers ebbed and flowed, carving landscapes and moving rock and earth along the way. As this theory of the ice ages gained traction, glaciers were thought to have slowly carved many familiar landscapes, from the Great Lakes and Cape Cod to alpine valleys and Scandinavian fiords and coastlines.

A "glacial pace" describes something that plods along, often painfully so. While the phrase can apply to a stultifying bureaucracy or long-awaited change, it does not describe the workings of glacial ice itself. Rignot and his colleagues revealed how ice the size of the North American continent can flow in rivers that move almost half a mile per year, often in ways that we do not fully understand. In fact, in the 1800s, scientists working on alpine glaciers found hints that ice can move and change shape quickly. One scientist noticed that the glacier that carried his research hut flowed downhill when he returned to make observations each year. Later, other scientists discovered that ice can flow almost as rapidly as seen in Rignot's satellite data by drilling stakes in the ice of the glaciers and following them year after year. The fact that ice can move astonishingly quickly has consequences for the natural world and the ways we have come to know it.

The following chapters show how moving ice shapes the world. As ice grows, shrinks, and flows, it creates new habitats for living things. When glaciers expand and contract, life adapts often in surprising ways—extreme environments bring remarkable new forms of life. Moreover, since the surface of the ice has captured objects inside for millennia, changes to glaciers expose the history of the cosmos itself. And, as we will see, rapidly moving ice has the power to change the planet on a time scale that can influence all our lives.

# CHAPTER 2

# HIDDEN WORLDS

On an early morning two months after the Nazis surrendered in World War II, U-boat 530 slipped into the Argentine naval base at Mar del Plata outside of Buenos Aires. One month later, another Nazi U-boat arrived at the station. Upon hearing word of this activity, Ladislas Szabó, a Hungarian exile, published an article promoting the theory that the submarines were secreting Hitler, Eva Braun, and Nazi leadership off to a hidden location in Antarctica. Szabó's idea gained a large following after he published a book on this idea of a clandestine Antarctic Nazi refuge.

The reality was far more mundane. When the U-boat crews were working off the coast of New York, they heard of the Nazi surrender. Wanting to escape to a potentially friendly country, they traveled on the ocean surface by night and under the water during the day to avoid detection. While their plan did not work, and they were immediately arrested, the myth kindled by Szabó's article on a secret Antarctic hideaway gained traction.

The idea tapped into a rich vein of fantasy that existed previously in science fiction. A few years before Szabó told his tale,

H. P. Lovecraft had published his classic horror novel *At the Mountains of Madness*, in which geologists from Arkham's Miskatonic University uncover alien civilizations under the ice. These kinds of fictional accounts and popular myths of subglacial military bases, alien spacecraft, and lost civilizations in Antarctica have been a common theme in popular culture. It turns out some wild ideas about what lies beneath Antarctic ice happen to be true.

Peter Kropotkin was born in 1842 into a Russian dynasty, inheriting the title of prince and a social network of important connections. He worked for the tsar in 1888 and later served in the military in Siberia. Kropotkin used his time in the service to explore his wide-ranging intellectual passions—politics, geography, geology, and biology. He devised theories about the ice ages by looking at Siberian landscapes. He later formulated new ideas of evolution that centered on cooperation and altruism as centerpieces for the changes in species. After his military stint, he became a different man. Kropotkin's ideas on the importance of cooperative behavior in biology led him to see government as a force that restrained human potential. To Kropotkin, the best form of government was as little government as possible. His political philosophy led him to decline the secretaryship of the Russian Geographical Society and relinquish his title as prince. Sent into exile, Kropotkin spent his days devising theories of nature and social systems that perpetuated his struggles against the tsarist government. At age forty, Kropotkin returned to Russia around the time of the Bolshevik Revolution, only to clash with that regime later.

During his exile, Kropotkin thought a lot about ice. He was impressed by the work of physicists such as Lord Kelvin in England, who speculated that ice under pressure can transform into liquid water deep inside glaciers. Kropotkin took this idea even further. Expeditions were returning from Antarctica revealing a continent of ice. Kropotkin saw something else. He proposed that underneath the ice of Antarctica sat vast lakes—entire worlds sealed off from Earth's surface.

Many people at the time thought Kropotkin's notion was no different from the fiction of Lovecraft or Szabó's alternate history. Kropotkin's theory of lakes under the ice sat in the realm of speculation, largely for the bizarre nature of the idea and the inability of the science of the time to test it.

In 1957, after the signing of the Antarctic Treaty, Russia built the first of its Antarctic research stations. One of them, named Vostok (Russian for "east") after the ship of the First Russian Antarctic Expedition in 1820, has the reputation of sitting on the most extreme and dangerous place on the continent, if not the entire planet. Located nearly 12,000 feet above sea level, on a distant part of the ice cap, Vostok has experienced the lowest temperatures ever recorded on planet Earth: –127 degrees Fahrenheit. The twenty-five crew, known as Vosties, are a legendary bunch who can tolerate deployments over a year long and isolation at Earth's most remote scientific base. Research at Vostok has not come without challenges. In 1959, the station was rocked when two Vosties got into a fight over a game of chess. One crew member attacked the other with an ice axe. While the fate of the victim remains obscure, there's a legend that the Antarctic Soviet banned chess for a period of time.

A unique kind of individual is drawn to work at Vostok. Andrei Kapitsa was a son of Russian scientific nobility, whose father won the Nobel Prize in Physics in 1978. Andrei developed a passion for Antarctica and did a PhD thesis speculating about the thickness and movements of the ice sheet. He noticed that aerial surveys of the ice around Vostok recorded something strange. Instead of having the normal corrugated topography of moving ice—with ridges, valleys, and hills—the ice around Vostok was entirely flat.

After his first Vostok experience, Kapitsa became interested in exploring inside the ice. In 1964 and 1966, he led a research team at Vostok to take seismic measurements of the movements of the ice below the station. By this time, seismic observations of the movement of sound waves through ice were revealing the internal properties of glaciers, such as their thickness and movements. But Kapitsa's readings were odd. Instead of seeing a signal for the thickness of ice and its contrast to the rock below, he was finding blurred interference, like static on an old TV. The readings made no sense. His research hit a stopping point. Then, most of his data were damaged in a fire in Moscow. As life moved on, Kapitsa forgot about his results.

By the early 1990s, Kapitsa had gained an international reputation for work at Vostok and was invited to a conference in England on polar ice. When British researchers presented the results of their aerial radar survey of the ice around Vostok, the patterns of their scans rang a bell for Kapitsa. He had seen these features before. The flight surveys suggested that the ice was not homogeneous, that it was dense on top with something much less dense below. Could this pattern be a sign of liquid water? Were the seismic readings, the topography of the ice, and the aerial radar soundings pointing to the same unexpected conclusion?

Returning to Moscow, Kapitsa was able to recover some of the records that had survived the fire. In assessing his earlier analysis of the seismic charts, he later recalled: "I looked at them and saw the mistake." He added, "When seismic waves go through water, longitudinal waves return and latitudinal waves do not. There were no latitudinal waves." Kapitsa's data revealed the presence of a lake underneath the ice, just as Kropotkin had predicted over a century before.

The ice at Vostok is almost 12,000 feet thick, and the lake sits almost two miles below the surface. Dubbed Lake Vostok, the sixth largest freshwater body in the world by volume, it is roughly the size of Lake Ontario. We now know that the lake contains several islands and is one of four hundred freshwater lakes under the Antarctic Ice Sheet. Some of them, such as Lake Vostok, may have been separated from the world above for over 15 million years.

In the late 1990s, inspired by the mysterious discovery of a lake under miles of ice, Russian, French, and American scientists set off to drill a core to get down to it. This project is arguably one of the most ambitious efforts ever in Antarctic research. Drilling a two-mile core is challenging anywhere. But Vostok is a different planetary environment altogether. The highest recorded temperature in summer is 6 degrees Fahrenheit, and there is scant time for a crew to use this "good" weather to ship and build the drill, and then to drill the core. Gear often fails at these temperatures, and resupplies are infrequent at best.

Coring into the Vostok ice comes with another serious complication. The lake below has been separated from the atmosphere above for millions of years. Any coring project would have to be mindful of potential contamination of this pristine environment. Drills can be messy things, as solvents, lubricants, and antifreeze compounds are required as part of their regular use. If released as a by-product of the drilling project, any of these could devastate the integrity of the waters of the lake.

Ice coring began at Vostok in 1970. Each season the team made some progress but then needed to depart as harsh winter weather set in. The story is a saga of trial and error. The first cores reached about halfway to the lake, roughly 6,000 feet, but then the drill bit got stuck in the borehole. The contraption needed to be hauled back up to the surface and repaired before drilling could start again. A series of holes were started over the next decade, only to have each fail for reasons of broken equipment or loss of the drill bit at depth.

Twenty years later, the core was widened and new technologies promised to finally allow the drill to reach two miles under the ice. But delays of supply flights and the challenges of drilling to these depths hampered the work. Because the team also was attempting to avoid drilling directly into the water below, they stopped the core about 400 feet above where they expected liquid water and pulled everything back up.

The ice at this great depth has special properties. The blue ice that is typical for these deep ice cores carries the chemical signature and physical structure of ancient snowfalls that became more compact and denser with each passing year. The ice right above Lake Vostok was different: it was far less dense and was composed of layers that looked like ice that typically freezes around large bodies of water. The ice had the texture and composition of what glaciologists call "accretion ice," which forms at the margins of lakes and oceans. This observation led to one possibility—this layer of ice reflected frozen waters from the lake below.

This core was humanity's first encounter with a lake that hasn't seen the surface of the planet since our ancestors had tails and were still living in the trees. The search for what lay inside became an unexpected new adventure.

U.S. scientists were given a piece of the accretion ice. John Priscu, from Montana State University, received a sample, about 20 inches long and 3 inches wide, and took it to the clean room in his lab in Bozeman, Montana. Putting the samples under high-powered microscopes, such as scanning and transmission electron microscopes, he screened for minerals, particles, and any other objects that could be trapped inside.

"And bingo, we started seeing microbes," he later said. When it rains, it pours—once he saw one cell, he started to see them everywhere. Counting the cells, he surmised that there were about 100,000 microbes per milliliter of ice, about one-tenth those in the ocean or in surface lakes. Then Priscu and his team set out to extract DNA from the water. The idea was to obtain sequences of all the DNA in the ice and compare them with databases of known microbial genes. By this method Priscu could rapidly obtain a picture of what kinds of life sat inside the sample.

DNA sequencing revealed *hundreds* of different species of microbes living miles under the ice. That piece of Vostok ice "changed my life," Priscu later said. He now had an entirely new research plan: he was going to hunt for life under the ice throughout Antarctica.

An amateur bush pilot, car racer, and scuba diver, John Priscu is a self-described adrenaline junkie. He also knows how to make bold plans. In 2000, soon after the publication of his Vostok results, Priscu was appointed to be the American representative for the Scientific Committee for Antarctic Research, a group that advises different nations about priorities for research on the continent. He was betting that his results definitively showed a huge diversity of life under the ice. Priscu pushed the committee to double down on that bet by looking for other places to drill for life under the ice.

Priscu's ideas weren't universally accepted. Some thought his results may have been artifacts of the drilling process. The ice could have been contaminated by the drilling machinery and the humans who touched it. Others felt that the lakes should be left alone, free of human influences and possible contamination.

Priscu spent six years planning his next steps. His approach was to take the path of least resistance. In Vostok, drilling takes decades and is fraught with potential contamination issues, logistic challenges, and political entanglements. Priscu's idea was to identify subglacial lakes that were not as deep as Lake Vostok, that were more easily accessible logistically, and where he could deploy sterile drilling technologies.

Subglacial Lake Whillans fit the bill. This lake sits 300 miles from the South Pole and is within a two-week tractor drive over the ice from McMurdo Station. Supplies could be ferried there much more easily than to Vostok. Moreover, radar and seismic studies revealed that Whillans is buried under only an estimated 1,800 feet of ice.

Using a drill sterilized by UV light and hydrogen peroxide to avoid contamination, Priscu's team was able to sample the waters of Lake Whillans. Back in the lab, Priscu did similar tests to those he'd performed on the Vostok core over a decade before.

DNA sequencing revealed nearly four thousand species living in Lake Whillans under the ice. Priscu's team was able to isolate and grow thirteen of them in the laboratory.

Far from being a derelict population of infirm stragglers, the subglacial microbes were diverse, thriving, and part of a complex web of ecological interactions. Being removed from the sun as a source of energy, these species evolved ways to extract energy by metabolizing the ammonium, sulfur, and iron in the lake waters. They evolved ways to share nutrients with one another during times of stress. Lying almost a micrometer from their neighbor, they even developed chemical signals to communicate as they shared their microbial food.

Radar and seismic data reveal that the waters of Lake Whillans flow like a delta into the Southern Ocean and have been connected to it for at least 100,000 years. There is an interplay between the life under the ice and that in the oceans: microbes and nutrients from Lake Whillans flow into the sea and influence the microbial community in the deep ocean waters.

In 2018, Priscu's team moved to a different subglacial lake, Lake Mercer. Lake Mercer is five times deeper than Whillans and reflects an environment farther from the ocean. The team deployed new equipment, including a sterile robotic probe to image Lake Mercer and a coring apparatus that could bring up the muck from the bottom of the lake.

DNA sequencing and microscopic analyses of Lake Mercer's waters revealed that its microbial life was less dense than that of Lake Whillans. But the mud from the lake bottom held some unexpected mysteries. The mud was returned to the lab at McMurdo, spread out on a slide, and put under a microscope. David Harwood, from the University of Nebraska, dryly described the contents of the slide as being "fully unexpected." He saw carcasses. Most were from single-celled foraminifera. But there were also shells of crustaceans, as well as a tardigrade, an animal the size of a grain of sand.

Tardigrades have eight legs and are also called "water bears" for their teddy bear–shaped bodies. They are known for their ability to survive in extreme environments and have been recovered from the sediments of the McMurdo Dry Valleys. They have the remarkable ability to shut their metabolism off and hibernate for almost five years. They can survive in places and conditions where other animals will perish: from extreme heat up to 304 degrees Fahrenheit to extraordinary cold down to –328 degrees Fahrenheit. They've even been shown to be able to survive in the vacuum of space. Because they accomplish all these feats while looking like cute teddy bears under the microscope, tardigrades have developed a loyal fan following.

Networks of flowing water under the ice may carry living creatures from lake to lake in this subglacial world. Another possibility is that the creatures in some places under the ice today have been isolated from the rest of life on our planet for thousands, if not millions, of years. While NASA sends probes to look for life on other planets, we still have places to search for life on our own. Only with more exploration of the worlds under Earth's ice will we know how life got there.

While Priscu was drilling his cores at Lake Whillans and making plans for Lake Mercer, the Russians were continuing with their own drilling at Vostok. In 2012, the Vostok team hit water. The hole had to be closed as the extreme weather was setting in and the team had to depart for the season.

Scott Rogers, a microbiologist at Bowling Green State University, obtained some Vostok ice from the core. Using newer DNA techniques and an updated microbial genetic database, Rogers made surprising, if controversial, discoveries. Like Priscu a decade before him, Rogers found microbes that can metabolize minerals in the dark environments under the ice. But there were other microbes that were far more puzzling. Several species he uncovered are only known to inhabit the guts of crustaceans like crayfish. Others only live in the digestive tracts of fish.

Some researchers are skeptical that there are either crayfish or fish living under the ice, let alone ones that have been isolated for millions of years. But who knows? The living world under the ice remains one of our greatest frontiers.

## Time Capsules

We were cracking jokes, pretending to be calm as our helicopter was being packed to take us to our first Antarctica field camp. Delayed by a week of bad weather, the team finally gathered at the helipad wearing the required uniform for a chopper flight in Antarctica: Big Red, a crash helmet, mountaineering boots, and snow pants. With the jitters at the start of six weeks in the field, and conjuring all the physical and mental acclimation that lay ahead, I became transfixed while watching the helicopter tech at work. She packed the Vietnam-era Bell 212 machine as if it were a three-dimensional game of Tetris, with gear of all shapes and sizes perfectly squeezed into every nook and cranny. Leaving just enough room for six humans in full extreme-weather gear, she placed us in the chopper based on our weight. As I boarded the craft, she smiled as she said, "You're going to the most beautiful part of Antarctica. Wait until you see Blood Falls."

As the craft lumbered over the Dry Valleys, moving ever closer to the Antarctic Ice Sheet, it was obvious that the only thing flying in this part of Antarctica was our helicopter—no birds or insects were in the air. No mammals, reptiles, or other creatures walked on the ground below either. The place seemed completely lifeless. The beauty of it lies in its desolation. Consisting solely of rock and ice, it seems as devoid of life as it does of any obvious human impact.

Beyond the Dry Valleys, a horizon of ice, with the occasional mountaintop poking through, extends as far as the eye can see. Flying in the midst of this sea of white, an isolated red island grew ever larger in the distance. Sitting like a red bull's-eye on the edge of a glacier, the shape of the feature emerged as we approached. One of the glaciers terminated in a cliff that rose about five stories high and stood out from all the rest for hundreds of miles. A river of red spewed out of the top, almost as if thousands of gallons of paint were gushing from its insides. Blood Falls was appropriately named.

The name reflects the history of the region. Thomas Griffith Taylor trained as a geographer in the early part of the twentieth century at a time when expeditions were launching to explore both the North and South Poles. He landed an extraordinary opportunity. Robert Falcon Scott was so impressed by Taylor that he appointed him senior geologist on his *Terra Nova* expedition in 1911 to attain the South Pole. Taylor was assigned the scientific goals of studying the ice, rocks, and animals of the region around the harbor in which the *Terra Nova* was at anchor. While Scott set off with five comrades in an ill-fated attempt for the Pole, Taylor and two others left on a mission to do geological investigations of the interior of the continent. Taylor's team found fossils, including the fish

Blood Falls

scale from ancient seas that led my colleagues and me to this region nearly a century later. His team also mapped the area and studied the geology of various rock formations.

Taylor was the first to see the red anomaly in the glacier during his trip in 1912. He proposed it was a patch of algae that somehow managed to find a foothold on the ice. He gave the strange feature the name Blood Falls and returned to the *Terra Nova* with his small team. Little did Taylor know that the leader of his expedition had already perished on his attempt to reach the South Pole. In fact, none of the crew had any knowledge of the tragedy until a search team was sent six months later.

Blood Falls is a target that has attracted scientists ever since Taylor's first encounter with it. In 2005, researchers scoured photographic records of the site taken by crews who had done flybys or fieldwork in the area. These archives reveal that Blood Falls acts intermittently—short periods of flow can be separated by long intervals of quiescence.

In the 1990s, scientists collected a sample of water from Blood Falls to look for the algae that Taylor supposed was inside. Nothing living was observed under the microscope. The red color is not coming from algae. The cause of the blood-red coloration lies in the chemistry of the water. The subglacial samples have a similar mineral composition to ocean water, except for two important differences. First, they contain a high concentration of iron. This fact explains the red color. The iron in the water is derived from the iron-rich rocks below the glacier. When the waters emerge from cracks in the ice, they interact with the oxygen in the atmosphere to make red rust. But something else inside the water was puzzling—it had extremely high concentrations of ocean salts. While the list of minerals in Blood Falls' water is the same as the list of those in the Ross Sea near McMurdo, Blood Falls' mineralization is more than twice as concentrated.

These discoveries led to further questions: Why does rust spill out of a relatively tiny patch in the vast Antarctic ice? And why would ultra-briny seawater be emerging from a glacier that sits almost a hundred miles from the nearest ocean?

The geology of the region holds the answers. There is widespread evidence that over 5 million years ago, the Southern Ocean extended deep inside the fiords in this part of Antarctica. Then, starting 5 million years ago, the climate became colder and glaciers encroached into the region. While the sea retreated, a portion of that ancient ocean water got trapped underneath the glacier as it grew. Under Blood Falls lies a fragment of sea that has been isolated from the ocean and sealed under the ice for at least 1 million years.

Lying between the ice on the surface and the rocks below, the waters of this dollop of ocean get saltier every passing year. As ice forms, salts remain in the ancient water that is still in the liquid phase. As more seawater freezes, the liquid water becomes ever brinier. This extraordinarily salty water lies in contact with the bedrock at the base of the glacier. As the salty water interacts with the brown rocks, it picks up the iron that gives the falls its distinctive coloration.

In 2013, a team of scientists ventured to the glacier now known as the Taylor Glacier to see inside using ice-penetrating radar. They deployed portable radar units over the glacier around Blood Falls. Shooting signals into the ice to measure the different densities of materials inside and to determine its contact with the rock below, they saw small channels within the ice where the briny water flowed. The glacier contains a network of these tributaries filled with ancient ocean water. The reason why water flows out of the glacier remains obscure. It may be that it sits at high pressure and gets squeezed out of holes and cracks in the ice. Another possibility is that movements of the ice make the water flow at some times and not others.

No life was initially visible in the briny water under the microscope. But by 2003 sequencing technology and computing power had gotten to the point where the waters could be explored in new ways. When the DNA in the water was extracted, sequenced, and compared with other living things, the conclusion was inescapable: life was inside.

While there were no algae, the water under the glacier contained the signature of over seventeen different kinds of bacteria. All of them are marine species, a fact that speaks to their life in an ancient ocean. The composition of their genes gives clues to their way of life. Like Lake Whillans' microbes, those in Blood Falls live with no sunlight, and found new ways to thrive. These creatures metabolize sulfates, common in the rocks below the glacier, for energy. They also have genes that promote survival in extraordinarily salty environments. Living in an ancient fragment of sea, removed from the sun's rays, these bacteria have been evolving in isolation from the rest of the planet and its life for over a million years. Along with the trapped ocean water that is their home, this community of microbes is a time capsule from a million years ago.

## Spilling Out and Spilling Over

Sitting 600 miles from the North Pole, Canada's Lake Hazen is one of the northernmost lakes in the world. With increased warming in the Arctic, meltwaters from local glaciers, streams, and rivers drain into the lake. In 2002, a team from Ottawa wanted to probe inside these waters to assess what is being released into the Arctic ecosystem as glacial runoff increases. Just like Priscu and Rogers did in the Antarctic, the Ottawa team collected water samples from the glacial runoff and sequenced the DNA and RNA inside. As expected, they found different kinds of bacteria. They tweaked their filters to expand their search and, almost immediately, viruses started to pop up. And not just a handful; there were likely hundreds of strains. Then the scientists compared the viruses' diversity to that of potential hosts in the area—the animals, plants, and fungi that live near the lake. Any similarities between the viruses and hosts could be a rough approximation of whether the viruses could infect the local fauna and flora. The result was unsettling. The team found that the viruses released by melting glaciers could spill over to local plant and animal hosts. Moreover, this relationship is directly tied to the amount of meltwater they sampled. This means that increasing glacial melt—that is, more warming—will lead to larger numbers of viruses being released that can infect local hosts.

An international team working on the Guliya Ice Cap in the Tibetan Plateau drilled into its ice and took samples back to the lab for DNA analysis. The genetic sequencing revealed that the ice held thirty-three species of viruses that had been buried for over 15,000 years. Four of the species were known, but twenty-eight were new to science. Each species has the genetic code that viruses use to infect microbes and plants. They appeared to have survived for such a long time because of adaptations in their DNA that favored an existence in the cold.

Everywhere we look for viruses in ice, we find them. As the polar regions melt, creatures from the past are exposed to the modern world. In 2014, researchers found thirteen viruses frozen in the ice of Siberian permafrost. Some of them had been trapped in the ice for almost 50,000 years. In the laboratory, the researchers isolated one dormant virus to test whether it could still function. They easily reanimated it and found this viral Rip van Winkle worked as if it had never slept for millennia. It was still able to infect single-celled amoebae.

There is a microbial double whammy to a warming Arctic. With increasing temperatures, new kinds of creatures can migrate north. Southern-adapted species will expand their range as they find new homes in areas that thaw. At the same time, viruses will reenter the ecosystem after having been frozen for years in the ice of glaciers and permafrost. This synergy can lead to escalating viral spillovers to local hosts—Arctic microbes, fungi, plants, and animals.

Newly emerging bacteria are part of this system too. In the summer of 2016, temperatures in western Siberia reached nearly 100 degrees Fahrenheit. The warm summers there in recent years have led to the melt of the upper layers of permafrost. Mud and sinkholes are starting to become regular features of the landscape. So, too, is animal disease. That summer, the nomadic peoples in the region noticed that reindeer were dying off in large numbers. By the end of August, more than 2,000 reindeer perished. In addition, 42 local villagers got sick that month and one of them, a twelve-year-old boy, died. The events were connected, as was revealed by tests of human and reindeer tissues: the telltale sign of anthrax infection lay in the blood and respiratory tissues of the victims. Anthrax can survive as bacterial spores that can lie dormant for decades. It is thought that these spores were released as the melting permafrost disgorged a long-dead infected deer from the ice. The deer was a victim of the last anthrax outbreak in the region seventy-five years ago. The spores from the newly exposed carcass then entered the water supply, where it later spread to both reindeer and humans. A plague from the past had reemerged in the warming permafrost.

Microbes reflect the endurance and resilience of life in polar environments. Anthrax and other microbes can survive extreme conditions by going into suspended animation for thousands of years. A clever adaptation enables bacteria to wait out harsh climates and lay low until favorable conditions return. These microbes have a protein that sits on the outside of the cell that is a chemical sensor for nutrients such as sugars. When these creatures encounter a nutrient molecule, even after decades or centuries, the sensor protein changes shape and initiates a chain reaction to spring the bacterium back to life.

Another biological strategy, microbial resilience, was revealed by a creative experiment. A team led by John Priscu asked the following question: What do microbes feed on when they live in lakes under the glaciers? Lacking sunlight or any obvious nutritional source under the ice, there seems to be nothing for them to metabolize. Priscu's team took some of the rocks dredged up by the cores of the subglacial Lake Whillans and pulverized them. They then let the rocks sit in water for over a month so that any chemicals inside could leach out. Looking at the water chemistry after the treatments, they found ammonia, carbon dioxide, methane, and hydrogen. A knowledge of bacterial metabolisms made this list of chemicals a gold mine of information—the mix is a smorgasbord for the microbes known to live in those waters. Some subglacial bacteria metabolize methane for energy, while others produce methane by metabolizing carbon dioxide. Still other microbes in subglacial environments metabolize ammonia. Yet another group of bacteria metabolize the sulfur that is produced when ammonia-eating microbes die, in essence feeding on their microbial corpses. A self-sustaining, thriving, and complex ecosystem can emerge from rock and water in the darkness under the ice.

The discovery of life under the ice in Antarctica now has scientists looking to the sky at the icy moons of Jupiter and Saturn. Both Europa and Enceladus are small moons covered by ice, and it's likely that there are oceans and rock below. On Earth, areas we thought were desolate are teeming with life. If life can thrive under the ice in Antarctica on eroded rock and water, perhaps it can do so in extraterrestrial environments. The work in Antarctica not only reveals how microbes can survive in surprising places, it prompts scientists to devise the technologies that could probe distant moons. Drills, drones, and laboratory protocols used for finding life under miles of Antarctic ice might be repurposed to explore Europa and Enceladus. Hidden worlds here on Earth may help us to understand alien life.

## CHAPTER 3

# COLD COMFORTS

"Don't even think of pitching it there." Bill, an intimating presence standing six-foot-three, with more summers on expedition in distant locales than I had years, was gesturing to a hairy black caterpillar on the ground below me. Bill was famously gruff, and this exhortation was likely the only thing he said to me the entire day. It was July 1988, and I was looking urgently for the flattest place to pitch my tent. My presumptive spot was adjacent to a tiny patch of tundra, an unusual sight within the bedrock, ice, and mud that formed most of the valley. Both puzzled by how a caterpillar could elicit such a reaction and humbled in front of the entire crew, I set off to find myself a new Arctic home.

In this part of Greenland, the landscape is composed of layers of brown sandstone that line a glacially fed valley. The hills are capped by snow and the cliffs erode to piles of boulders and rock that carpet the valley floor. Clays between the rock layers break down as water flows over and through them. The resulting mud accumulates on flat surfaces and can swallow boulders and, as I was to learn, boots and legs up to the thigh.

These rocks hold 200-million-year-old fossils from the Triassic Period: dinosaurs, turtles, and large amphibians—all animals adapted to tropical environments. But here at 78 degrees North Latitude, temperatures can get as high as 40 degrees Fahrenheit during the twenty-four hours of daylight in summer and drop to –20 Fahrenheit in the dark of winter. Like Antarctica, the region is a polar desert, with annual precipitation less than 7 inches. The glacier that lies at the head of the valley dominates much of the local environment. The area can generate strong winds, and when the sun is bright, the winds turn the dry valley into a raging torrent of meltwater and mud that can last for a week or more.

The High Arctic doesn't seem particularly suitable for abundant plant, animal, and fungal life. The lack of greenery, the diminished energy coming from the sun, and the fierce cold in winter demand special adaptations of any creatures that dwell here. While microbes are everywhere in these polar landscapes, larger animals are not easily seen.

The longer I spend in this landscape, the more its living world comes into focus. Some of the most productive nature expeditions result from putting my head right at ground level.

In low-lying places flat enough for mud and sediments to accumulate sits the Arctic's sole species of tree, the Arctic willow. This willow, with true branches, vascular systems, and leaves, does not grow upward toward the sky; it grows sideways, along and under the ground. Its trunk and branches rarely rise higher than a few inches into the air. The roots and branches follow cracks in the rock, spaces between rubble, or patches of mud. If there are few obstructions, the branches might extend as far as 30 feet. The leaves are green until the end of July, but in August they turn orange, a clear indicator that the Arctic fall has arrived. The willow's slow growth is coupled with its longevity. One specimen in Greenland was found to be over two hundred years old.

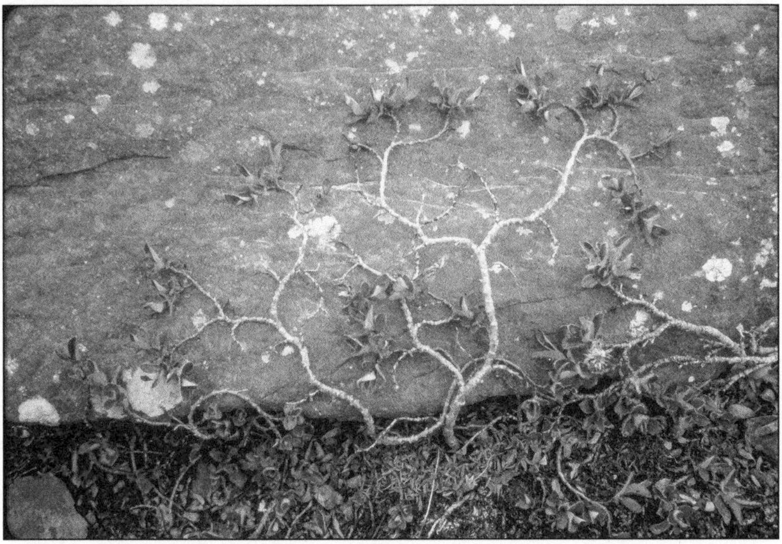

An Arctic willow growing sideways along the ground

The Arctic willow is well adapted for life in polar valleys. Its shallow roots obtain nutrients above the impenetrable permafrost layer. Small hairs on its surface protect its leaves and reproductive structures from wind and cold. Growing like a carpet along the Arctic floor, it is protected from fierce winds.

Where there's Arctic willow, mosses and short grasses form a spongy mat of tundra. Occasionally, wildflowers—in a range of yellows, purples, and blues—rise less than five inches high. Tiny mushrooms, the size of a dime, sprout from the moss's surface. In some parts of the Arctic, mostly farther to the south, the tundra can extend for miles. Where I've hunted for fossils, however, tundra is limited to small patches that fill crevices in the rocks and rubble of the valley floor.

This world at ground level is far more nurturing than even a foot above. In 2002 our camp was hit by a large gale with sustained winds over 70 miles per hour and gusts even higher. At the height of the storm, two of our tents began to shred on their windward sides. For several hours we scrambled among the tents in raging winds to secure them and the gear inside. We managed to move between the tents by taking oblique angles to avoid walking directly into the teeth of the winds. On occasion, surprisingly large gusts came out of nowhere and threw all four of the crew to the ground simultaneously—nature's way of telling us to be quadrupedal rather than bipedal. Fortunately, the ground was soft with mud and patches of tundra. Being on the ground during the storm was a revelation; it felt like an oasis from the ferocity above. At the heart of the gale, this place was quiet and still. Here, life continued uninterrupted. Spiders crawled along the mosses, bees flew low among wildflowers, and caterpillars fed on other low-lying plants. This windless world is also warm because it radiates the heat absorbed during twenty-four hours of sunlight in the summer. Low elevation forms a microhabitat for animals, plants, and fungi to thrive even in the harshest of conditions.

The caterpillar that caused me such public embarrassment lives on these surfaces too. The Arctic woolly bear caterpillar reveals the complex mix of adaptations required to survive in the far north. As the name implies, it is a caterpillar about an inch long and covered with bushy hairs that extend in all directions from its body. The Arctic species has cousins that live farther south in the continental United States. Seen throughout North America, the woolly bear caterpillar has gained a popular myth about its coloration. The southern species are banded in black and brown, and their color in summer has been said to be a premonition of the coming winter—the wider the black band, the harsher the winter will be. While the folklore has never been verified, the caterpillar's Arctic relatives live a life that is even more worthy of myth and legend.

The woolly bear is only active for the growing season in the High Arctic—an interval that covers the month of July. The caterpillar emerges in June, immediately after the snowmelt, and begins

Woolly bear caterpillar on the tundra

actively feeding on its food of choice—the leaves of the Arctic willow. On sunny days, it will take a chance with the predators flying above and bask in the sun to warm itself while the hairs on its body retain heat inside. The woolly bear feeds voraciously on the willow, using the short period of relative warmth to stock up on energy. Then as mid-July arrives, the willow starts to lose its leaves and the woolly bear builds a silk nest in the ground to protect itself from predators.

As temperatures begin to drop, the woolly bear does its biological magic—it freezes to become a solid hairy cylinder.

When temperatures rise and snow melts the following June, the woolly bear thaws and emerges to begin feeding, basking in the sun's heat, and storing energy. The cycle repeats when the woolly bear freezes again in the fall. Ecologists who have looked at the growth rates of the woolly bear estimate that it lives in this freeze-thaw cycle for at least seven years, and possibly as many as fifteen.

Each year the woolly bear stores ever more energy, until the summer it has accumulated enough resources to transform into a moth. The moth flies for two weeks, finds a mate, expends its accumulated energy stores, and dies. Nearly a decade of freezing and thawing, feeding, and basking, all the while avoiding predators, is all in the service of two weeks of flying and mating.

This remarkable life cycle is why the caterpillar in my desired camp spot elicited such a strong reaction from Bill. I was about to injure (or worse) a living thing that likely had endured years of polar life. What's more, the flat tundra surface that was so attractive for a tent is also the likely home for even more caterpillars and other creatures.

Without the ability to freeze like an ice cube, the woolly bear caterpillar's lifestyle would be impossible. But freezing is mostly devastating for cells and tissues. When tissue freezes, the water inside it forms ice. Ice has a greater volume than water, so as it expands, cells burst. Almost as bad as freezing is thawing. When frozen tissues melt, cells deform and their fluids can tear through the membranes around the cells. The thaw is the most damaging part of frostbite in humans for this reason.

The woolly bear has a convenient trick to survive freezing: it makes its own antifreeze. With this trait it becomes an elegant example of what evolution can do—the woolly bear can withstand temperatures as low as –70 degrees Fahrenheit.

The woolly bear caterpillar is a case study in polar adaptation: it has a life cycle tuned to the light and dark seasons of the High Arctic, insulation to preserve its heat, feeding strategies and storage abilities to hold nutrients for at least seven years, and "antifreeze" proteins that allow it to survive freezing solid. These strategies are employed in different ways in diverse creatures in both southern and northern polar regions.

When bodies get cold, surprising things can happen. The experience of Anna Bågenholm is a case in point. Bågenholm was training to be a physician in Norway and was an avid skier. In 1999, she was out with two friends and lost control of her skis and fell headfirst into freezing ice. Only her feet and skis were poking out of the ice. Her friends worked desperately to extract her, but she was wedged in firmly. Bågenholm found a small pocket of air to breathe while her friends worked to get her out. As they waited for help, Bågenholm's body went limp and her heart stopped. After eighty minutes, the rescue team finally extracted her and found her pupils dilated, skin white, and pulse still absent. They medevacked her to the hospital in Tromsö, but the electrocardiogram on the flight showed a flat line. The hospital staff couldn't find a pulse either, and their EKG also showed no heart activity. Knowing the body's responses to the cold, they hooked Anna up to a heart-lung machine to give her oxygen and gradually warm her blood. After one day, her heart was able to beat on its own. After twelve days, she opened her eyes. It took another year for her to be able to move because of damage to her nerves. Bågenholm is now a practicing radiologist working in the same hospital that revived her.

The lead doctor on Bågenholm's case later quoted a famous truism about hypothermia: "You're not dead until you are warm and dead." Lower temperatures mean the body consumes less oxygen. The brain, a huge energy user, can survive for longer periods without circulation at colder temperatures than at warmer ones.

A breakthrough in understanding how animals can thrive in extreme cold came from Arthur DeVries of the University of Illinois, working in McMurdo Station in the 1960s. DeVries observed that polar fish survive in waters that are below normal freezing temperatures. The salt in the ocean waters surrounding Antarctica causes ice to form at lower temperatures than the standard 32 degrees Fahrenheit. DeVries studied icefish, a species that thrives in the cold waters of the Southern Ocean. A Norwegian physiologist observing their behavior in the 1950s had hypothesized that polar fish must have some sort of antifreeze in their blood. So DeVries set off to find it. Analyzing their blood chemistry, he was able to isolate a protein that had the right atomic structure to do the job—it could bind to ice crystals to stop them from growing. As the fish's body gets ever colder, ice crystals begin to form but are inhibited from growing large enough to damage the cells.

The antifreeze story got more interesting when DeVries and his colleagues sequenced the protein and compared it with other proteins. It was almost identical to a digestive protein in the fish's liver. With a few small mutations over time, this liver protein was repurposed into an antifreeze in the ancestors of icefish.

Since DeVries's initial discovery, antifreeze compounds have been discovered in different fish, insects, plants, and fungi living in both the northern and southern polar regions. The structures of the molecules in the species vary, but the biological effect is the same: the molecules either prevent ice crystals from forming or they limit the damage to cells by the freezing and thawing process.

Industries have taken an interest in these natural antifreeze proteins. Imagine having an ice cream that doesn't form ice crystals after months in a freezer. Unilever attempted to manufacture just that in 2007. They took the gene that makes the antifreeze protein in an Antarctic eel and genetically modified it to make it in industrial quantities to put into their ice cream, code named "VanEELa." Obviously, the marketing department did not coin the moniker and the product never made it to the shelves.

Antifreeze is not the only means to avoid the cellular damage of extreme cold and ice-crystal formation. Another mechanism is to just get rid of the water in the body altogether—with no water, there is no freeze-thaw damage. Many invertebrate animals desiccate in the extreme cold and dry conditions of Antarctica. The molecule behind this process is a sugar known as trehalose. When roundworms start to dry out, their cells begin to make the sugar. As trehalose accumulates in each cell, it replaces the cell's water, and the creature becomes a dry husk as its metabolic processes shut off. The sugar mimics water in its atomic reactions and keeps the proteins in the cell from being damaged. In some cases, so much sugar accumulates in the body that the entire organism transforms into a brittle substance that can shatter—a sugary glass. The worms can survive in this glassy state for years until they become reanimated by the addition of water. Some worms have been reanimated after decades spent in glass form.

This strategy of trading water for sugar, and occasionally transforming into glass, can be seen in different polar organisms—animals such as worms, tardigrades, and small crustaceans, as well as fungi and single-celled organisms. Trehalose is a general-purpose tool for the biological strategy of desiccation and has evolved independently in each species. Creatures that do not use trehalose sometimes form a protective shell around their bodies and retreat into a cyst during harsh conditions.

Creatures with insulation don't need antifreeze proteins or to go into suspended animation as glass or cysts. Fur coats on mammals, and downy feathers on birds, can take a variety of shapes. Musk oxen are a common sight in July because they are feeding voraciously, rarely even taking notice of nearby humans. Musk oxen look, as the name implies, like large bovine creatures with a coat of hair that extends almost to the ground. In reality, their closest relatives are not oxen, but goats. Shave a musk ox and you see a naked goat with shorter legs and broader horns. The coat of a musk ox contains two layers of wool, both of which trap air and hold body heat inside. The outer layer is thick and more permanent year to year. The inner layer, known as the qiviut, is such a good insulator that it is highly prized by Alaskan ranchers and Inuit hunters. It is stronger and warmer than sheep's wool, and softer than cashmere. In July, when the temperatures on Ellesmere Island rise above freezing, the musk oxen shed their qiviut, and you can find it on the leaves of Arctic willows, rocks—pretty much everything musk oxen rub against.

Polar bears also have two layers of hair, an outer coarse layer and an inner fine one. Polar bear hair is not actually white, even though it may look that way. Their skin is black and absorbs the heat of the sun. The outer hairs are hollow and completely transparent. The apparent whiteness is due to the way in which the hair scatters light. It serves as a prism that releases all spectra of light, giving the appearance of a white coat. Fly tiers have prized it for its ability to float and transmit light in tantalizing ways to attract trout.

Not all parts of an animal's body can be covered by layers of fur or feathers. The feet of penguins are completely exposed to the elements and contact the ice and cold rock directly. The legs of seabirds in polar regions extend into frigid waters that are often below standard freezing temperatures. These creatures use their circulatory systems to retain heat inside the body. The arteries that run from the legs to the feet are bundled in parallel to the veins that return the blood to the body. The heat in the arteries is picked up by the colder veins that return blood to the animal's core. This means that the warmth of the blood in the arteries stays in the body and does not dissipate into the environment. With this neat trick, little heat is lost.

Body shape, too, serves as a kind of insulation. There's a mathematical formula that governs the shapes of bodies that work best in polar regions. Because exposed surfaces are places where heat can be lost, animals need to minimize their surface area relative to their body weight. Mathematically, this is described as a surface-area-to-mass relationship where surface area is proportional to the square of the length of the animal and mass is proportional to the length cubed. As length increases, the value that is cubed will grow faster than the value that is squared. This relation means that heavier and larger animals will have less surface area than smaller ones. Biologists have quantified this proportion and noticed that polar species, on average, are relatively larger than their southern cousins. A corollary to this rule is that the longer the limbs, the greater the potential for heat loss. Limbs project from the body and are exposed to the elements. Longer limbs mean an animal has a greater surface area than one with shorter limbs. Consequently, polar animals tend to have shorter appendages than those that live closer to the equator.

Behavioral adaptations work with anatomical ones to conserve heat. Many animals, such as polar bears, can roll their bodies into a ball to reduce their surface area. Others, such as musk oxen and emperor penguins, take a group approach: a herd or group will huddle together and share each other's warmth. Still other creatures can retreat to burrows in the tundra or spaces between rocks.

The rhythms of a polar day can be fodder for strange encounters. One night at two a.m., nature called and I had to begin the unpleasant process of getting myself out of a cozy sleeping bag, and a deep sleep, to venture into the cold. Aided by the fact that it was as bright as day at this hour, I was admiring the view when I heard a series of grunts behind me. Wheeling around, I saw I was in the presence of a large male musk ox about twenty feet away. Normally, proximity to a lone male would be cause for panic, but this animal was busy. He was feeding so intently on willows at the margin of camp that he either didn't notice me or didn't care. He was hungry and had to consume as much food as possible. This strategy, feeding in the middle of the night—seen in many polar creatures in summer—requires unique adaptations.

Being at the poles means adjusting to the rhythms of twenty-four hours of daylight in summer and twenty-four hours of darkness in winter. Summer is the time to feed and put on fat to store enough energy for the lean times of the polar winter. Herbivores, such as musk ox, reindeer, and others, need to feed continuously during the short summer season to store enough food to survive the bleak winter. Twenty-four hours of daylight poses special problems for animal bodies. Animals from flies to humans share a fundamental trait: molecular and genetic clocks in their bodies that are tuned to a twenty-four-hour day and cycles of light and dark.

The Nobel Prize in 2017 was awarded to three scientists who deciphered the molecular machinery that drives the clock inside cells. Working on fruit flies, and doing years of elegant experiments, they showed that protein levels in cells rise and fall during the twenty-four-hour day in a set rhythm. These diurnal molecular changes lead cells to behave differently throughout the day and night. This clock works with receptors inside our brains that take the light from our eyes to send signals to different parts of the body. Consequently, all bodily functions—digestion, hormone secretion, the action of the kidneys, and body temperature—follow a rhythm over the course of the day.

Virtually all animals, including humans, are finely tuned to the spinning of the Earth and the light and dark cycles that result from it. We all know the effects of jet lag, but there are other profound effects from changing our internal clocks. Disruption of circadian rhythms can be detrimental to health. For example, humans who work a graveyard shift and are exposed to artificial light from midnight to eight a.m. are at greater risk for certain cancers. Presumably the propensity to develop cancer is related to the fact that DNA copying happens on a diurnal cycle. When this process is disrupted by a change in the circadian rhythm, errors and mutations that cause disease can appear.

The same holds true for other animals. Yet many polar creatures can feed continuously throughout the day in the summer, giving up sleep altogether with no ill effects. In these cases, their body clocks are either disabled by genetic mutations or they run in a twenty-four-hour cycle but their effects are disconnected from the animals' physiology. This ability to disconnect from internal clocks is a rare ability among living things.

Polar ecosystems are local and patchy. Looking for fossils in polar terrain means being sensitive to changes in texture and color of the landscape. Fossils typically stand out because they are a different tint or shape from the rocks we typically encounter. In Greenland, Ellesmere Island, and Antarctica, the rocks of interest tend to be brown or red, much like those of the deserts of the American Southwest. But brown rocks and mud aren't the only colors we encounter when doing Arctic hikes. Occasionally we'll see a bright-green surface in the distance. The rarity of these sightings becomes a source of attraction—we are pulled to the green spots like iron filings to a magnet.

These isolated islands of green often hold a large surprise. Rising like a natural tombstone from the green carpet below, a pair of musk-ox horns projects through the moss. The space around the horns is a vibrant garden of mosses, lichen, and wildflowers. Small white and yellow flowers, mushrooms, and insects, including woolly bear caterpillars, move through these lush islands. If this space is on a slope, the green carpet extends downhill for a short distance with bones of the musk ox strewn within. Often, an entire musk-ox carcass lies just underground.

These green habitats sprout from the nutrients that were released by the decomposition of the musk ox's body over many years. Once you know how to look, you can see these natural musk-ox graves in nooks and crannies throughout the landscape. A musk ox spent its life ranging over miles of the Arctic, eating willows and other small plants. Now those nutrients are concentrated in a small patch that supports life. A musk ox's carcass transforms a nutrient-poor mud into a Garden of Eden teeming with living things.

A musk-ox garden: a decaying body becomes an oasis of life.

Any sliver of landscape can become an oasis for life if it concentrates nutrients. In 2004, just above our camp on Ellesmere Island, we saw a boulder the size of a kitchen refrigerator. It seemed strangely out of place—covered by orange and green lichen, it stood out like a fluorescent warning sign on an interstate highway. We dubbed it "lichen rock," and it became a familiar welcome on our treks home from long hikes. Surrounding lichen rock was an intimate ecosystem of tundra, insects, and plants. But there was something else contributing to the life in this small patch of the valley. Bird dung covered much of the rock and tundra surrounding it, and tiny skeletons of rodents set in pellets excreted by the birds littered the ground. Birds, particularly jaegers, made this their home. Jaegers are famously aggressive, and if we got too close to lichen rock they would dive-bomb us, coming a foot or so from our faces.

A sequence of events makes a barren sliver of Arctic land a thriving community of living things. The large rock in the center of the valley attracted birds of prey seeking perches. The ejected carcasses of their prey and their bodily waste fell to the ground, injecting nutrients into the previously inhospitable muds. The rich nutrients allow lichen, tundra mosses, willows, and other life to sprout. Their life cycles, in turn, make this small patch ever more fertile. An unusually tall boulder, isolated in a broad valley, became a venue for life, death, and renewal.

Any small feature in this nutrient-poor landscape—the carcass of a dead animal, a tiny depression that would allow particles of animal waste to accumulate, a rock that could serve as a perch for birds—can become an island of life.

Animals, plants, and fungi have been evolving for millions of years to adapt to the extremes of light and dark, cold and wind, and the dry desert conditions in polar environments. Years of natural selection have endowed these creatures with tricks to live in these harsh landscapes. The human story, however, is more complicated.

## Fish out of Water

Many of our uniquely human characteristics came about as adaptations to the warm, subtropical environments of East African rift valleys. In that region, our human ancestors developed a nomadic, bipedal lifestyle. Large brains supported increasing cognitive abilities with ever more complex tools and, likely, social interactions. The problem is this: our brains only function within a narrow window of temperatures. If human bodies get too hot, our brains shut down. Accordingly, managing heat in the savannah was the name of the game for our ancestors. To stay cool, they lost a hairy coat of insulation and evolved an enormous number of sweat glands in the skin—more than any other mammal. Given a history of adaptation to dissipate heat, human bodies seem uniquely ill-equipped to manage the cold and dry climates of polar regions.

Since the origin of our species 300,000 years ago, humans have moved from being tropical and temperate animals to polar ones. Human history near the poles begins in Siberia 40,000 years ago. Later, around 4,000 years ago, Inuit traditions and technologies allowed populations to live nomadic existences in the far north. Inuit Peoples developed new ways of making shelters, tools for fishing and hunting, and clothing from local materials: ice, snow, pelts, and bones. These populations developed high-fat and high-energy diets along with cultural practices that allowed them to dwell in cold and dark conditions. Through this time, as with all human populations, their bodies and genes changed as they adapted to local climates and diets. For some human populations, adaptation to animal domestication led to the ability to digest lactose. Human adaptation to the polar world involved a different set of genetic changes.

Inuit health has long been a paradox for Western medical researchers. By necessity, the traditional diet is low in fruits and vegetables and high in fats derived from seals, whales, and other marine animals. Given their fatty diet it would be expected that Inuit would suffer from cardiac conditions, such as atherosclerosis, that cause suffering in people farther south. But that is not the case. Inuit heart health exceeds that of most European, Han Chinese, and North American populations that have been studied.

In a search for answers to this puzzle in 2015, scientists from the University of California, Berkeley, and the University of Copenhagen compared proteins in the blood of Greenlandic Inuit populations to those in Europeans and Han Chinese. The team found an extremely significant difference—100 percent of the Greenlandic people tested had a combination of genetic variants that were found in only 2 percent of Europeans and Han Chinese. These genetic adaptations play a very specific role in the body—they are involved in the metabolism of the kinds of fats found in marine mammals. Known as desaturases, these enzymes convert saturated fats to unsaturated ones. As we know from trips to the doctor's office, a diet rich in unsaturated fats is more heart-healthy than one heavy in saturated ones because of the effects on other lipids. The genetic variation seen in Greenlandic Inuit naturally promote heart health; they work to lower blood levels of the "bad" cholesterol (LDL) and raise levels for the "good" cholesterol (HDL) in the bloodstream. These enzymes also help to lower fasting insulin levels in the blood. These effects are critical in preventing the formation of dangerous atherosclerotic plaques in blood vessels. Because of these genetic changes, Greenlandic Inuit have healthy hearts even while regularly consuming extremely rich fats.

Each of us, whether living at the poles or not, is able to adapt to cold. Some effects are immediate. The dry and cold environments of the Arctic and Antarctica affect the entire body. We have specialized nerve endings in our skin that are sensitive to temperature. When the temperature of the skin starts to drop, signals sent to the brain set off reactions in different parts of our body. Cold temperatures initially cause the body to produce heat through shivering. When we shiver, our muscles contract involuntarily. With each twitch, heat is released. While shivering is an involuntary response, people working in the cold are taught to do jumping jacks, abdominal crunches, and lunges as ways to intentionally generate more heat. I'll routinely do abdominal crunches when I get into my cold sleeping bag before going to sleep each night. The burst of activity makes for a cozy furnace inside.

The cold triggers nerve signals to the lining of our blood vessels. These nerves cause small muscles in the walls of the arteries to contract, shunting the flow of blood from the periphery to our core.

Since heat is lost at our skin, the change in the way our blood flows keeps our vital organs functioning while conserving heat.

Many people from more temperate zones who work in dry polar environments deal with an unfortunate side effect of cold adaptation. When cold weather shunts blood from the periphery to the core, more blood is delivered to the kidneys. The surging volume of blood is linked to the need many people feel in cold regions to urinate frequently. Polar explorers and mountain climbers have commented on this inconvenient effect, particularly when bundled in parkas and snow pants covering onesie underwear in subzero temperatures. The challenge is that these are dry places too, so the water lost to urination and breathing needs to be continually replaced by hydrating regularly.

In most other mammals, cold temperatures cause small muscles inside the skin to make hair in the fur stand erect. That's why your pet might look bigger and fluffier when outdoors in the winter, like they just had a bad hair blowout. Humans have this muscular response in the skin too. But because we lack extensive hair, we just get goose bumps.

Fat is useful tissue in polar regions because it serves both as an insulator and also as a storehouse for energy. Humans produce at least three kinds of fat tissue: white, brown, and beige. Fat molecules, present in each of these kinds of tissue, are poor conductors of heat. This fact makes them great insulators. Fat tissue contains different kinds of cells: some structurally hold up the fat molecule, whereas others perform the metabolic functions necessary for the tissue to work properly. As adults, we have mostly white fat, which lies under our skin and around our organs. Brown fat is common in infants and is seen in isolated patches in adults. Unlike white fat, brown fat has cells that contain numerous mitochondria, tiny organelles that burn energy and generate heat. This means that brown fat is not an inert tissue. With the ability to store energy, insulate the body, and also generate heat, brown fat can be a special adaptation for those who live and work in cold regions. Beige fat, as the color implies, is intermediate between the other two kinds of fat, both in terms of color and in the number of mitochondria found in the cells.

While our bodies respond to cold immediately, some internal reactions take weeks or months to develop. In 2014, researchers at Virginia Commonwealth University and the Garvan Institute in Australia took five volunteers for an experiment that involved a serious commitment. Each of the five was signed on to a four-month gig in which their lives were highly controlled. They were given a diet in which every nutrient was counted and consumption measured. Each month they received physiological tests of energy exertion, blood work to assess sugar levels, and MRI scans of their entire body. And here is the hard part: they were given a bedroom with only a bedsheet to cover them at night, while the researchers tweaked the temperature every month. At first the space would be at room temperature, then the temperature would go down to 64 degrees Fahrenheit, then it would be warmer in the following months.

The results showed the ways in which human metabolism can change with temperature. After the month of chillier nights, each of the volunteers developed 42 percent more brown fat than they had the previous month. Then, at the end of the experiment, their percentage of brown fat decreased to the levels they had at the start of the experiment.

These experimental results are in line with observations of human populations—people who experience cold tend to have a greater percentage of brown fat than those who do not. In fact, recent work suggests that cells in our body respond to different temperatures to produce different types of fat. Cold can directly induce stem cells throughout our bodies to make more brown fat than white.

Acclimation to the polar environment takes many forms. Whenever I first arrive onto a field site from the warm confines of the base, I'm perpetually cold. My initial indication of acclimation comes when I'm hungry morning, noon, and night. I'll even wake in the middle of the night ravenously hungry. After a week in the field, I begin to tolerate lower temperatures. I start needing fewer and fewer layers. The opposite is also true: returning to base after living in a cold tent can be sweltering.

There's both observational and experimental evidence behind this kind of adaptation. At five foot four, with a beak of a nose that gave him his nickname, Henry Robertson "Birdie" Bowers, was one of Robert Falcon Scott's most cheerful and hardy crew members. He was the storekeeper for the *Terra Nova,* and no task was too onerous for him. According to the memoirs of the expedition, Bowers never had bad days—rarely without a smile, he was ever cheerful regardless of the external conditions. He was famous among the crew for his daily routine. After rising, Bowers would pour a bucket of ice water over his head. His daily ablution wasn't only about being clean, it was derived from his theory about acclimating to the cold. Bowers hypothesized that regular, short exposure to cold helped him adjust to Antarctic life.

Recent experiments suggest that Bowers's frigid routine was no mere boast—we can become desensitized to the cold. When experimental volunteers are subjected to different regimens of cold, whether being put in a cold pool for a period or in a cold room, their responses will change. Over time, people acclimating to the cold shiver less and their vessels do not shunt blood as much as before at the same frigid temperature.

The altered daily cycle of light and dark at the poles wreaks havoc on human bodies and minds. Studies of scientists at polar bases during the dark of winter show that their sleep cycles, hormone levels, and energy can be skewed. In a polar base such as

McMurdo, artificial light, enforced mealtimes, and work schedules give the residents a sense of a daily rhythm, a clock to which their bodies can become in sync. Despite this consistency, people's internal body clocks can go awry with altered sleep and activity cycles. It's not unusual to need a nap during midday or to wake at two a.m. ready for work.

The most extreme example was seen in a study of a four-person crew who overwintered in a small hut in Antarctica. Their physiological states were monitored twenty-four hours a day for the entire period they were in the dark. Lacking the cues of work schedules, mealtimes, and extensive artificial lighting, their body clocks went, as circadian biologists call it, "free running." Their hormone levels and sleep cycles functioned on an internal clock independent of the planetary one. Most of them had a daily sleep-wake cycle of more than twenty-four hours. They ended up going to sleep later and later each day. So, if their normal bedtime was 10:00 p.m., they fell asleep at 10:10 the next day, 10:20 the day following, and so on. The time set by their body clocks became disconnected from those of the outside world. After eight weeks, their internal body clocks shifted to the point where their bedtime was early morning.

One of the greatest difficulties for outsiders on polar expeditions has been staying sane. With the physical challenges caused by extremes of temperatures, and with light cycles that can wreak havoc on bodies and moods, the most successful expedition leaders have had to learn how to keep morale high and their crew mentally stable. The renowned, if controversial, polar explorer Dr. Frederick Cook famously said of the situation on his ship in Antarctica, "Here men love and hate each other in a passion which defies description. Murder, suicide, starvation, insanity, icy death, and all the acts of the devil, become regular mental pictures."

For those of us accustomed to lower latitudes it is easy to become untethered from reality in the strange and isolated environments at the poles. Expeditions working both in Antarctica and the Arctic have had to deal with crews who became unhinged. One Argentine crew member burned down his Antarctic base after learning he had to overwinter there. And one Russian stabbed his Antarctic colleague for spoiling the endings of books. Then there's the Australian who became so dangerous he had to be locked in the freezer of Mawson Station in the 1950s. There was also the physician who poisoned his own patient and expedition leader, Charles Hall, on an American expedition to the North Pole in the 1800s. Tales of depression, psychosis, mutiny, and murder are rife in accounts of exploration at both poles.

Tolstoy famously opened *Anna Karenina* with the classic line "All happy families are the same; every unhappy family is unhappy in its own way." The statement is as true for polar work as it is for family life. Expeditions that have minimized mental suffering and promoted well-being have typically followed certain practices. Lessons learned from these efforts have stood the test of time—they are as relevant on scientific bases and deep-field camps today as they were aboard ships in the nineteenth century.

John, one of the Australians on our Antarctic expeditions, was a veteran among workers on the continent. Part of Margaret Bradshaw's Kiwi teams that mapped the area in the 1980s, he had access to the New Zealand polar station adjacent to McMurdo. As we were stuck on base for a few weeks, John arranged a tour and dinner at the station as a fun diversion. McMurdo felt like bustling Manhattan compared to the New Zealand station's seeming tranquil Vermont village—it is smaller and more intimate than McMurdo and studded with displays of the history of explorers from Edmund Hillary to Margaret Bradshaw. Toward the end of the tour, our host took us to a large room with a dramatic view of the Ross Ice Shelf. In McMurdo, a similar room houses the scientific library. At the New Zealand station, however, the room was lined with racks holding costumes, mostly plus-sized women's dresses and gowns. I later discovered that McMurdo once had a similar costume room at some point early in its history.

When living in a confined space with twenty-four hours of daylight or dark, thousands of miles from home, in an extreme climate, entertainment comes at a premium. Explorers in the nineteenth century knew that shows, many of them in drag, were a way to lampoon personalities and hardships, form a bond around common themes, and forge a community. Lacking cues to time, days in the polar calendar can merge into one, and celebrations can put stable markers on the monotony. Dramatic performances, newsletters, and special menus can break dreary cycles.

Many expeditions celebrated holidays with special meals and printed menus giving fancy names to regular polar fare, with some special treats added that were brought along and stored away for the occasion. The Greeley Expedition's attempt to attain the North Pole in 1891 followed this tradition with their Christmas dinner. Their menu read "Mock-turtle soup, salmon, fricasseed guillemot, spiced musk-ox tongue, crab-salad, roast beef, eider-ducks, tenderloin of musk-ox, potatoes, asparagus, green corn, green peas, cocoanut-pie, jelly-cake, plum-pudding with wine-sauce, several kinds of ice-cream, grapes, cherries, pineapples, dates, figs, nuts, candies, coffee, chocolate." While the Greeley Expedition's desserts were provided by one of their generous New York patrons, our Christmas treats in Antarctica over a century later are from Trader Joe's. Our well-traveled grocery-store cargo includes chocolates, cookies, and tins of cake.

Successful expeditions have put sharing meals at the center of group life. Everybody has comforting foods and smells that recall hearth and home. For Americans, morale builders can be as simple as mashed potatoes, mac and cheese, or pizza. When interviewing potential team members, I probe them about their favorite foods and dietary restrictions. The range of comforts for an international team can be as diverse and simple as a bag of lentils, green-tea matcha Twix bars, or Vegemite.

Arguably, the most essential part of stressful work in uncertain conditions is ensuring that people feel valued, heard, and that their well-being is a priority for the expedition. Ernest Shackleton was affectionately nicknamed "The Boss" by his crew, because they knew he would suffer any privation for their safety and for the success of their group effort. And he did just that when he rescued the crew of the *Endurance* in 1916 from South Georgia Island.

Engaging people in decisions, including the most junior members of the expedition, gives them agency that can help fight hopelessness, despair, and homesickness. Giving the team chances to celebrate individuals' special abilities, whether drawing, Thai cooking, or storytelling, is a way to break up the monotony and build group cohesion. And most important on a small team, knowing when *not* to talk is a critical skill. It's often best to leave certain things unsaid or, more commonly, let people have their personal space. Experienced leaders and team members either have a knack for this, learned by example, or have come to understand it the hard way.

## Rules of Working in Polar Regions

Extreme environments force us to see our bodies as machines. We must become hyperaware of and proactive about our levels of hydration, heat retention, and energy consumption in a way that is often lost in the comforts of a more temperate home. People who move to work in the Arctic and Antarctica follow rules for surviving, even enjoying, these places. While the directives may seem intuitive, it might surprise you to learn how many of us have accidentally violated them from time to time.

**Rule 1: Don't get cold.** It is easier, and less risky, to keep our bodies warm than it is to heat them up. If you need to warm up, then use your body as a machine. In Antarctica, jumping jacks or sit-ups can turn Big Red into a personal furnace. But beware Rule 2 below.

**Rule 2: Don't get hot.** Getting wet can be a great relief in tropical environments, but it can be fatal in polar ones. Whether you're trudging through snow or climbing a hill, it is easy to start sweating. Perspiration, our response to heat, chills the body through evaporation. On traverse, I try to be mindful of my heart rate and how much my body is heating up. If I can't chat and tell tall tales on a hike, then I'm likely overexerting myself, and going to sweat too much. Thermoregulation breaks are important when on long traverses, not only to catch one's breath but to keep from sweating. I'll often stop to remove layers, only to put them on again fifteen minutes later, then repeat the process all over again to maintain the right balance. The general advice for working in cold places is: rush only when necessary. Most tasks can be done at a thoughtfully slow pace to avoid injury, making bad decisions, or overheating. Productive hiking and working in these places entails finding a Goldilocks zone consisting of a comfortable heart rate, manageable heat balance, and effectiveness on the job.

What's more, layers of clothing must follow specific patterns. The closest layer to your body should wick away perspiration. Cotton is particularly bad at diffusing sweat;

wool and various synthetics are ideal. On top of this layer, middle ones hold insulation, while the outermost layers protect from wind and moisture.

**Rule 3: Don't get dry.** It is easy to dehydrate in cold environments because the cold does not make you get thirsty. When on the terrain, I bring two thermal flasks with me. One bottle holds hot water, the other starts the day empty. I'll fill the empty bottle with handfuls of snow throughout the day. By adding dashes of hot water to the snow, I can keep a room-temperature supply of water going for ten hours.

Lips, faces, and hands can peel; noses can bleed; and eyes can get too dry in the wind and cold. To make matters worse, sunburn can happen even inside a nylon tent. The inside and outside of our bodies need continual hydration and protection in these landscapes.

**Rule 4: Keep your head in the game.** Attitude is everything; it is important to keep a light touch, even a sense of humor, about yourself and toward others. Holding a feeling of gratitude, even under harsh conditions, works wonders to elevate the spirit. Focusing on the science we're there to do, the natural attractions in the landscape, good books, and the talents of teammates is a recipe for maintaining a level head in the face of challenging circumstances.

Just like in camp, food is not only fuel when on traverse, it is comfort for the soul. I'm always nibbling chocolate or dried fruit. Bring goodies.

**Rule 5: It takes a village.** Inuit Qaujimajatuqangit, which translates to "that which Inuit have known to be true," refers to traditional knowledge that guides interactions with the land

and with one another. Essential matters like working together, care for one's community, and concern for others become a time-tested social glue. Practicing this traditional wisdom promotes wellness for individuals and the community at large. Versions of these ideas, either gained by working with Inuit communities or developed independently, have also guided outsiders on successful polar expeditions. In Antarctica, for example, we are always doing buddy checks of each other's mood, attitude, alertness, and well-being. This is not a place to work alone, and the person suffering from a field ailment is often the last one to know it.

**Rule 6: Think inside the box.** Stay within your limits, do not take risks, and when a situation feels out of your control, focus on the things that you can influence. Thinking "outside the box" is for the whiteboard at home, not for polar expeditions.

**Rule 7: Experts mess up.** Veterans of polar work can sometimes be nonchalant about procedures or overconfident about dangers. It's important to stay vigilant and not allow other members' expertise or experience lull you into becoming recklessly passive. Hence, Rule 5 only works if everybody's opinion and observations are taken seriously.

Science at the poles comes with three priorities in rank order: ensure each other's safety, do the work we are there to do, and have the time of our lives. Promoting well-being, the obvious highest concern, means connecting to approaches used by Indigenous Communities and explorers of the past, and it means staying humble with the knowledge that each of us raised in more equatorial climes is a fish out of water in polar landscapes.

# CHAPTER 4

# OUT OF THIS WORLD

Building 157 of McMurdo Station sits at the margin of the Ross Ice Shelf and is an anomaly on base. It has no windows and sits separate from the well-trodden walkways that connect other buildings. Only rarely have I seen people enter and exit the building. The running joke among polar researchers is that this is the place where the alien bodies are stored, a nod to horror movies and legends of UFOs buried under Antarctic ice.

The reality is that Antarctica is in fact covered with aliens. You just have to know how to look.

At the turn of the nineteenth century, polar exploration was the equivalent of the space program of the twentieth. Ships crammed with the highest technological gear were launched by militaries and private individuals to probe vast unknown regions. As the race to the poles ignited national fervor, diverse personalities were pulled into the quest, from sailors and machinists to doctors and scientists. A successful trip to the poles was a ticket to adventure, discovery, and possible fame.

Among those drawn to Antarctica in the early twentieth century was the Australian Douglas Mawson. Mawson trained as an engineer in Sydney and shifted to geology to combine his twin loves of travel and exploration. After early success mapping the rocks of Melanesia, he caught the fever for work in Antarctica—here was a place he was going to make his mark. Mawson was recommended by a senior colleague to be one of the Australians to join the legendary British explorer Ernest Shackleton on his first (unsuccessful) quest for the South Pole. Mawson distinguished himself on this trip—Shackleton's *Nimrod* Expedition in 1907 to the Ross Sea—by being among the first to scale the local volcano, Mt. Erebus, and trek to the magnetic south pole. He so impressed Shackleton that he was invited to join Robert Falcon Scott in 1910 on his *Terra Nova* Expedition in Scott's attempt for the South Pole. But Mawson declined and hatched other plans.

Douglas Mawson in a moment of rest during his Antarctic expedition

In 1912, when Scott and Amundsen were sitting in their respective Antarctic harbors to attempt to gain the South Pole, the former tragically, Mawson launched an Australian expedition of Antarctica on the Adélie Coast, directly to the south of Australia. Mawson's goal wasn't to head to the South Pole but to explore the region and map glaciers, rocks, and coastline. Mawson had big plans for a modern expedition: a radio station would communicate back to Australia in a first attempt at direct communication between the Antarctic continent and elsewhere, while a monoplane would perform the first aerial reconnaissance of the continent. He recruited a young British jack-of-all-trades, Frank Bickerton, to maintain the gear. Unfortunately, the monoplane crashed in a test run before heading to Antarctica, and its wings were destroyed. Bickerton proposed that the wingless plane with its single propeller would make an excellent "air tractor" with skis. Instead of flying, it could be hitched to sledges to pull them along the ice.

Mawson developed the ambitious goal to cover as much of Antarctica as possible in the one year scheduled for the expedition. His idea was to divide and conquer by setting up multiple three-man teams to explore different parts of the region with sledges or the air tractor. The groups would travel hundreds of miles over three months and return to the base to share their results.

Mawson's team, including two other young explorers, Xavier Mertz and Lieutenant Belgrave Ninnis, set off with sledges pulled by dogs. This was challenging work because this part of Antarctica is known for near constant winds that can average 80 miles per hour, blowing on top of glaciers with an imposing number of deep crevasses. Mawson and Mertz each stood on their sledges as the dogs pulled them, while Ninnis, an enthusiastic fitness buff, ran alongside his dog-propelled sledge.

After five weeks, and a journey of almost 300 miles from the base, Mertz looked up from his sledge and saw only Mawson and his dogs. Ninnis wasn't anywhere to be seen in the vastness of the white landscape. Calling for Mawson, Mertz scanned the area to no avail. Approaching the location where they last saw Ninnis and his sledge, they discovered a giant crevasse. Some 165 feet into the crevasse were two dogs—one dead, the other gravely wounded. Mawson and Mertz dropped lines down into the crevasse to find Ninnis and rescue the surviving dog, to no avail. Ninnis was never seen again. The crevasse was hidden by a weak snow bridge, and Ninnis, because he was jogging, did not disperse his weight as well as a sledge with runners. Unfortunately, as the bridge buckled, the team lost vital supplies. Ninnis's sledge contained virtually all their food, their tent, and the fittest dogs. Lacking basic survival gear, Mawson and Mertz performed a quick funeral ceremony by the mouth of the crevasse and turned around immediately to make the treacherous 300-mile sledge back to safety.

With only a week's supply of food and no tent—but plenty of fuel and a stove—Mawson and Mertz took to twenty-seven-hour intervals of continuous sledging in an effort to get home as quickly as possible. They improvised a tent from a sledge cover sheet and their skis. Lacking any other source of food, they sacrificed dogs regularly, leaving nothing to waste. But on the trip back the two explorers became feeble and delirious, forgetting equipment and mis-tying knots.

Their condition became worse with every passing day—their skin began to shed, their nails became weak, and they suffered from nausea and cramps. The two started to have fits of delirium and insomnia, making a bad situation even worse. One night Mawson had to sit on Mertz's chest to keep him from destroying the tent during a bout of hallucinations. Mertz soon died, leaving a weakened Mawson with a solo trek of 100 miles to the safety of his colleagues back at base. Mawson trudged home in his frail state, even falling into a crevasse at one point. The width of his sledge kept him from descending all the way and he somehow found the energy to pull himself out.

When Mawson stumbled back into the company of his crew, he was delirious, despondent from the loss of his two comrades, and near death. His situation got worse from there.

A few hours before Mawson staggered into base, his ship, and most of his crew, had departed for Australia. Mawson could even see the silhouette of the craft sailing away in the distance. Six men were left to overwinter in the hopes that Mawson and his two companions would return. Among those waiting was Frank Bickerton, who held some welcome news for Mawson. Given Mawson's dire circumstances, Bickerton had to wait a few weeks for the expedition leader to recover to the point where he could appreciate it.

While Mawson, Mertz, and Ninnis were on their ill-fated sledging trip, Bickerton had led his own three-man group with the newly converted air tractor. The team did not make it far. The engine of the wingless monoplane failed in the extreme cold, despite Bickerton's constant fiddling with its inner workings. Jettisoning the plane and lacking the expedition's dogs that Mawson's team took with them, Bickerton's team had to pull the sledges themselves over hundreds of miles of ice. Theirs was a traverse through complete whiteness as they pushed into the interior with hundreds of miles of ice around them. They made small progress each day.

On December 5, 1912, the whiteness of the landscape was broken by a small black object, no bigger than a softball. It lay in a small depression in the ice and looked as if somebody had thrown it down on the surface. It had a black, shiny crust around it that appeared to have been blistered with fragments lying scattered about. Bickerton had a hunch that the object was important, so he took detailed notes and collected the specimen to bring to Mawson, who, as an experienced geologist, could confirm the discovery.

> We picked up a meteorite approximately 5" 3" 3.5" covered with a black scale, internally of a crystalise [*sic*] structure, most of the surface rounded except in one place which looks like a fracture, iron is evidently present in it. It was lying about 2.5" below the mean surface and did not appear to have been there long, probably only a month or so.

Mawson was near death when he encountered Bickerton upon his return. What had killed Mertz and nearly claimed Mawson was the dog meat. They had eaten every soft part of the dogs, including the liver. The livers of the huskies were so enriched in vitamin A that both men suffered from vitamin A poisoning. Mertz did not like the tough muscle meat and ate more of the liver than Mawson did. Hence the different fates of the two men.

After Mawson recovered, he was able to confirm that Bickerton's rock was, indeed, a meteorite—the first meteorite discovered in Antarctica. While they had to wait almost another year for a rescue ship, Mawson's expedition was a scientific and geographic success despite the enormous human cost.

It took another forty years for the next meteorite to be discovered in Antarctica. A Russian team working in the mountains of East Antarctica, nearly the other side of the continent from Mawson's expedition, picked up another rock with a black outer shell.

The meteorite story took a new turn when in 1969 a retired Japanese professor was giving his blessing to a team of scientists who were about to set off to study the glaciers near the Yamato Mountains in East Antarctica. The group consisted of ten Japanese geologists, cartographers, and glaciologists aiming to map the ice as Mawson's team had attempted fifty years earlier. Before departing, the leader of the group, a young scientist named Masaru Yoshida, made a pilgrimage to visit Professor Masao Gorai, an éminence grise in Japanese Antarctic geology, for his advice and good wishes for the trip. Yoshida asked Gorai if he would like a gift of any research samples from Antarctica that the team could collect during their work on the ice. Professor Gorai said, "I am not currently interested in metamorphic rocks from Antarctica; instead, please bring me a gift of ultramafic rocks and/or meteorites." Being a junior scientist starting his career and knowing that a total of only four meteorites had been discovered in Antarctica at the time, Yoshida set off with no small amount of trepidation.

Once on the ice in Antarctica, the team traversed it to set up stadia rods in different places to triangulate ice movements of various parts of the glacier. The work proceeded apace until one of the team members saw a shiny black rock set in a depression of the ice, much like the one Bickerton had encountered over fifty years earlier. It had a black crust, small pieces of which were left shattered on the ice. Thinking it a potential meteorite, Yoshida collected it to give to Professor Gorai. But then a couple days later the team found another strange rock. And then another. The following weeks became as much about finding black rocks on the ice as they were about mapping the glacier. That year, Yoshida's crew returned to Japan with nine dark rocks with glassy shells. Suspecting them of being meteorites, they wrapped each of them carefully for the return trip to Professor Gorai.

Expecting a triumphant welcome to Gorai's house with his nine potential meteorites, Yoshida discovered the professor was away. Yoshida left the box containing the samples with Professor Gorai's wife along with a short note describing the collection. When Professor Gorai returned, he didn't expect anything important in the box and did not open it. When Yoshida visited later, he reminded Professor Gorai of the box and the two opened it and unwrapped each of the specimens.

After laying the rocks out on a table, Professor Gorai looked

The Japanese team on their 1969–1970 meteorite hunt in the Yamato Mountains of East Antarctica

down with some confusion. These specimens were, as he said, "getemono." One English translation of this Japanese word is "offbeat, often referring to disgusting food most people won't eat."

A few days later, Professor Gorai set off to work in his lab on the rocks. The standard protocol is to slice a section, less than an inch thick, to observe the minerals and grains under the microscope. Each of the rocks had a dark rind on the outside that looked like melted rock. Meteorite experts call this layer a fusion crust, referring to the ways the heat of entry into Earth's atmosphere can fuse minerals together to make a shiny outer shell. Observing these details, and the strange combination of minerals inside, Gorai saw a whole new world inside the Antarctic collection. He fired off a telegram to Yoshida: "The 'getemonos' have all been found to be meteorites: all of them!"

Then the situation got strange. The more Gorai looked at his sections under the microscope, the more he saw that the combination of grains inside each of them was different—each was composed of a distinct array of minerals. If a large meteor had broken up during its passage through the atmosphere, the nine rocks should have had similar mineral compositions. But each of the meteorites was entirely unlike the others.

The conclusion, while completely unexpected, was inescapable: This meteorite collection wasn't of a single large meteorite that broke up as it fell. It was made by nine separate meteorite falls.

Was there something special about the hills in the Yamato Mountains? Did more meteorites fall over Antarctica than elsewhere on Earth? Or was there something about Antarctica that concentrated them?

The Japanese team had a hunch: it was the ice. As we've known from the days of Rendu, Forbes, and Kelvin, ice moves, folds, and bends. Perhaps there was something about moving ice that concentrated the meteorites that fell on it? The Japanese team proposed that the blue ice of Antarctica was special, that its movements caused meteorites that landed on it to be preserved on and in it, and that in some places the movements of the ice concentrated them.

The team assembled their ideas to present to the 1973 Congress of the Meteoritical Society in Davos, Switzerland. In the audience was Bill Cassidy of the University of Pittsburgh. By that point Cassidy had had a long career in geology, training as a geochemist specializing in the formation of meteor craters around the world. He had traversed the globe to assess the ways that meteor impacts affected the Earth's crust. His early training took him to Australia, where he met Douglas Mawson. While an interest in craters led him to the Davos meeting, his encounter with Mawson sat deep inside him. Cassidy came to the talk not because he was particularly interested in meteorites but because Antarctica had long captured his imagination.

Cassidy's life was changed by the Japanese team's presentation. So, too, was our understanding of the solar system. As Cassidy describes in his memoir, a lightbulb went off in his head during the talk: "Meteorites are concentrated on the ice in Antarctica!"

Immediately after hearing the talk, Cassidy hatched a plan. His idea was to collaborate with the Japanese scientists to launch surveys of the ice in Antarctica to explore why the ice concentrates meteorites. Then, using these ideas, he hoped to find places across the continent where he could gather a representative collection of meteorites. This was the age when expensive Apollo missions were bringing rocks back from the moon. Cassidy's notion was to check the blue ice in Antarctica for pieces of the solar system that had been delivered to Earth for free.

Cassidy applied to the National Science Foundation for funding. He was rejected. He didn't give up, and after three failed attempts, he was funded and ready to launch in 1976.

Cassidy established a profoundly altruistic approach to work on Antarctic meteorites. He realized that a collection of Antarctic meteorites could set up a gold rush, where different teams would work competitively and specimens would be separated into hundreds of different collections. Cassidy also realized that if the whole business was led and run by a single exclusive club, then scientific discovery would suffer because new people and ideas would be constrained from participating and entering the field. To avoid these problems, Cassidy made sure the composition of the search team each year wasn't limited to his students, his friends, or their students. Instead, each year he offered an open application for a team member slot.

This selection process allowed a new generation of meteorite students and researchers to be trained. Moreover, access to the meteorites by global researchers, the lifeblood of discovery, would be entirely free. To keep out cronyism, the people who made the decisions about access gave up their rights to the samples discovered. As the current leader of the Antarctic meteorite hunt,

Bill Cassidy on the ice

Professor Ralph Harvey of Case Western Reserve University, told me: "This is planetary exploration, done cheap, and crowdsourced. In my humble opinion, there's never been a boon to science that compares."

It's hard to disagree with Harvey's assessment. The program Cassidy launched, the Antarctic Search for Meteorites (known by the acronym ANSMET), has been running continuously since 1976, training hundreds of researchers.

More than 50,000 meteorites have been recovered as of last count.

## The Sky on Ice

In December 2018, I was getting antsy from our nearly three-week stay in McMurdo. Our training was finished, the crevasse drop was completed, and we were sitting in our small cubicles practicing tying knots and memorizing the procedures for crevasse rescue. I'd catch myself peering out my small window at the mountains and ice in the distance. Other teams came and went into the field while we were waiting for our slot. As they say, the closer you get to something, the longer the wait.

The Crary Building is the home for science in McMurdo. It contains the fume hoods, lab benches, and chemical storerooms found in any university lab. The safety measures here are far more intense, however, as is logical given the setting. It is stunning to look around this lab, about twice the size of a high school gym, and realize that every single part of it—from the doors to the chairs and chemicals—had to be transported to Antarctica on a ship that can only arrive during a short season each year.

The lab is as much a gathering place for stories of expeditions as it is a venue for doing experiments. There's a shared knowledge among Antarctic researchers that is necessary for doing work on the ice. I found myself whiling away the wait to go into the field by learning about what scientists had planned for their study of glaciers, the Dry Valleys, and the South Pole. It is quite easy to hear the inside scoop about other expeditions because the walls of McMurdo are thin. Sometimes I'd be privy to the challenges, both personal and scientific, that my colleagues faced.

One day, the conference area was occupied by ten people meticulously going over the details of their expedition to be launched in a week's time. About eight students and junior scientists were gathered around the table while a mountaineer and seasoned field leader were laying out the details of their expedition plans. I was working on an environmental safety report, one of several we'd have to do during the season, and was aching for a diversion. This departure was welcome—here were ANSMET meteorite hunters doing their preparations.

I poked my head out of the office and the team leader gave me the nod that it was OK to listen in. He seemed to be in his mid-sixties with long gray hair. The wrinkles above his beard were earned from decades spent on glaciers and in deep field camps. As he gestured over the Antarctic maps on the table in front of him, he had the low-key combination of calm and humility of somebody who has seen the best and the worst that polar regions can offer. The ANSMET team's plan was to take one of the large LC-130s, loaded with snowmobiles, sledges, and gear for a month, to mountains 400 miles away. There, they would launch smaller reconnaissance missions looking for meteorites on the ice surface. Whenever they found meteorites, they would GPS the site and collect them using tongs, sample bags, and other gear specialized for their preservation. The team was diverse in age, experience, and background, just as Bill Cassidy had envisioned forty-five years before.

Forrest came up and elbowed me from behind. "That's Johnny Alpine," he said, indicating the team leader. "If you want to learn about meteorites, he's your guy."

Following the presentation, I introduced myself to Johnny, or more properly, John Schutt. Like many polar scientists, Schutt was drawn to the ice by the twin loves of geology and mountaineering. After college, his gigs as a skiing instructor and mountain guide gave him contacts to join the crew on an Arctic base set up on a large moving iceberg known as Fletcher's Ice Island in Alaska. But his dream was to work in Antarctica. After a few failed attempts, he eventually got lucky. A mountaineer friend was scheduled to guide a bunch of scientists on a meteorite hunt, but since he was getting married that year, he wondered if Schutt could join in his stead. Schutt called Bill Cassidy to offer his services and heard nothing for months.

By that time, Schutt had become renowned for daring rescues

on Denali in Alaska and on other North American peaks. In one incident, he was roped to a client as they were traversing a knife-edge in the Saint Elias Mountains in British Columbia. The edge was narrow and bordered by slippery ice with a thousand-foot drop on each side. When Schutt's client slipped and fell over one side—a near-certain fatal plunge to the bottom—Schutt didn't hesitate. As he saw his roped partner slipping to one side of the knife-edge, he dove over the other side to brace the fall. When Cassidy heard that story, he hired Schutt to guide his team of meteorite hunters. By the time I met Schutt in 2018, he was over seventy years old and had guided the ANSMET team safely to far-flung and remote mountains in Antarctica for the past forty years.

I told Johnny we were headed to Deception Glacier, and his eyes lit up. "That's high on our list, it's loaded with blue ice. Look for meteorites when you're out there."

That suggestion was like catnip to me. Being a fossil hunter looking for ancient bones standing out in the background of the rock, I felt a natural kinship to the meteorite hunters looking for black rocks on the surface of blue ice.

Johnny's presentation led me to ask the question: What about blue ice makes it special for the collection of meteorites? Cassidy and his team had spent the first two decades of their meteorite program trying to find answers. The main insight was not only that ice moves—it's that it moves in predictable ways.

Mathematical models of the Earth's gravitation and surface, and years of data collection from around the world, tell us that the rates of meteor falls are similar across the planet. Meteors land at the same rate in Antarctica as they do in Africa, North America, and everywhere else. The difference is that Antarctic meteorites travel after they hit the surface, because they land on ice. In the early days of his meteorite program, Cassidy used to hold "rock races." Rocks of different sizes were left on the ice and their movements were

An Antarctic meteorite emerging from the ice

marked from year to year. The movements are slow in the places Cassidy worked, only a few feet per year. Extrapolating over long periods of time, the rocks could move across the continent. In other places, the movement of ice is even orders of magnitude faster.

When a meteorite hits the ice of Antarctica, it gets buried by falling snow. Because the interior of the continent is so cold, snow from each year buries it ever deeper and in some cases the meteorite gets entombed in blue ice, the densest part. Then the ice takes its inexorable path to the coast, carrying buried meteorites with it. Because of this movement, most of the meteorites that land on the Antarctic ice end up in the ocean, never to be recovered. If this were the complete state of affairs, we'd be witnessing a tragedy for science.

The good news is that Antarctica is a continent with land below miles of ice. In some places, mountains poke through the ice. Here, the flow of the ice from the center of the continent to the coast is interrupted. As the ice flows toward a mountain, the rock serves as a barrier. The ice buckles like a factory conveyor belt hitting a wall: whatever is on that conveyor belt, whether pies or boxes, will accumulate against the wall. As we saw earlier, this movement means that the deepest layers, the blue ice, get extruded to the surface. Meteorites trapped in the snow and blue ice pile up at the base of these obstructions.

The equation discovered by the Japanese researchers, Cassidy, and his team is a simple one: ice movement + barriers = the potential to find meteorites.

Most meteorites are not uniform but hold inclusions of various minerals, some smaller than a sesame seed. These inclusions are readily visible under the microscope in rock sections of the kind Professor Gorai made. While minerals can be analyzed with microscopes, determining their atomic composition requires high-energy devices that can separate the atoms and measure their masses. Some of these atoms are libraries of the ancient history of meteorites.

One of the most informative atoms is a version of aluminum, known as aluminum 26. Meteorites arrive on Earth carrying a parcel of it. When rocks—such as asteroids, comets, and smaller fragments—orbit around the sun, they are bombarded by cosmic rays. The energy of these rays is enough to transform elements inside the rock into aluminum 26. But aluminum 26 is radioactive, so the aluminum 26 inside the orbiting rock decays into another element. This process—whereby aluminum 26 in an orbiting body is generated by cosmic rays, only to be lost by radioactive decay—determines the level of aluminum 26 inside a meteorite relative to other elements.

Here's the key point: When a meteorite lands on Earth, our planet's atmosphere and magnetic field shield it from cosmic rays. Aluminum 26 is no longer produced inside the meteorite, and it starts to decay. With production of aluminum 26 stopping the minute the meteorite hits the ice, a recently landed meteorite will have an increased amount of aluminum 26 relative to those that landed in the distant past. The aluminum 26 in older meteorites has had a longer time to decay than in those that just landed. With these insights, scientists can use the relative amount of aluminum 26 to infer how long the meteorite has been on our planet.

Frank Bickerton found a rock on the ice in 1912 and guessed it was a meteorite that hit the Earth nearly sixty years before. He was judging the amount of time the meteorite had been there by the shape of the depression it made in the ice. Of course, he had no idea of the ways in which ice moves. Aluminum 26 tells a different story. Some Antarctic meteorites have been on Earth for 30,000 years and others are over 300,000 years old. There are a few meteorites that are estimated to have landed on Earth over 2 million years ago. The movement of the ice has taken eons to bring these ancient extraterrestrial visitors to the surface.

When meteorites hit the Earth, they, like every other rock exposed on the surface, begin to erode by the action of wind and water. In most places on Earth, meteorites weather away within hundreds or thousands of years, depending on the action of these forces in the local area. Antarctica, being a dry, cold desert, preserves these ancient visitors. And finding them gives us a record of the movements of the ice and the origins of our world.

## Treasure Hunt

All six of us on the fossil-hunting team were inspired to add meteorites to our search for fish. We gathered with Johnny around a large table in the conference area for a short briefing on the types of rocks we were most likely to see. Johnny broke out a weather-beaten loose-leaf binder and opened the faded blue cover gingerly to keep it from ripping off of the book's spine. Yellowed acetate held pictures of the kinds of meteorites that the ANSMET team had discovered since 1976. Here was a book that had seen decades of Antarctic storms and had been on blue ice around the continent. Stories of exploration—polar storms, long treks, and exciting discoveries—lay within each crease, fold, and weathered surface of that binder.

The most common kind of meteorite, one that makes up 94 percent of the entire collection, is of the type that Bickerton picked up in 1912. The black fusion crust surrounds a grayish interior composed of grains the size of beach sand. Between the grains are small inclusions, known as chondrules, much like those Gorai analyzed under his microscope. These meteorites, known as chondrites, can be found on every continent on Earth.

Atomic dating suggests these kinds of meteorites are 4.56 billion years old—almost a million years older than Earth. The grains and chondrules appear relatively undeformed by heat or pressure, as only the external surfaces are blistered and melted by their entry into the atmosphere. These small meteorites hold the material that swirled around the sun prior to the formation of the planets and asteroids. The components of the meteorites reflect the brick and mortar of our part of the solar system, the basic ingredients that went into the making of our planet and others. Ancient ice on Earth has coughed up some of the most primordial parts of our galaxy.

Another kind of meteorite, less common than chondrites, looks like a deformed hunk of metal. I naïvely thought these would be relatively easy to see on ice, as there isn't a lot of metal ore lying around. These meteorites look like chunks of Swiss cheese that has been pulled apart; holes and pores penetrate their surfaces and their edges are rough. Their shapes and elemental composition, mostly of iron, make them look as if they have been through high temperatures and pressures. As we've known since Charles Laplace in the 1700s, as planets and asteroids form, their cosmic dust and debris orbits the sun and starts to form ever larger clumps. Gravity pulls these clumps together and they grow and occasionally collide into one another like billiard balls. The bigger they get, the hotter their core becomes. The heat causes the metals in the interior to separate out. It is thought that these metallic meteorites are pieces of these cores that were broken away during collisions and that then entered Earth's orbit.

The third, very rare, class of meteorites are known as planetaries. The name suggests their origin. The crystals inside them contain stories of Antarctic and space exploration.

## Enigmas on Ice

As legend holds, at nine a.m. on June 28, 1911, in the village of El Nakhla El Baharia, Egypt, about 20 miles east of Alexandria, a dog was hit by a meteorite and vaporized into ashes. At the same time, villagers heard a loud boom and saw a smoke trail and a series of flashes. When the dust cleared, about forty stones, weighing about 22 pounds in total, were found strewn across the town. Some of the stones were buried 3 feet underground. These were odd rocks with large crystals inside. And several fragments were encased in a black, shiny crust, just like the fragments Frank Bickerton saw on the ice in Antarctica.

Some of the rocks from El Nakhla El Baharia were gifted to the British Museum and the Smithsonian Institution. When scientists got their hands on the meteorites, they discovered that they were identical to other strange ones that had fallen over Nigeria, France, and India decades before. These rocks stood out from other meteorites collected at the time. They looked similar to igneous rocks that appeared to be produced in lava flows. But where did cooled lava come from in space?

The mystery deepened even as new technologies allowed scientists to assess the atoms inside. Virtually all the meteorites discovered to this point—from Antarctica and across the globe—dated to the beginning of the solar system, about 4.56 billion years ago. But the specimens from Egypt and elsewhere were infants in comparison—their ages came in at about 1.3 billion years. Where could such young rocks have came from? All the known sources for meteorites—cosmic debris, asteroids, and comets—were too old.

Perhaps they came from the moon? When moon rocks were brought to Earth by the Apollo missions, they were analyzed geochemically. The ratios of different oxygen atoms in moon rocks and those of the crust of the Earth were shown to be identical. When data from oxygen in the meteorites came in, things got even stranger: their ratios of oxygen atoms revealed that they were not even part of the Earth-Moon system.

A relatively young rock from space, formed from cooled lava that didn't come from the Earth-Moon system, was puzzling. These fragments were too young to have come from asteroids and comets. That left few alternatives. Did they come from one of the other planets, like Mercury, Venus, or Mars? If so, how?

In 1979, Bill Cassidy led the ANSMET team to a small mountain on the Antarctic ice called Elephant Nunatak ("nunatak" is the geographic term for a mountain that pokes through the ice). Cassidy was drawn to this area by aerial photography done by the U.S. Geological Survey, which suggested it was potentially rich in blue ice. From the air, the dark rock looked like the trunk, hump, and eye of an elephant, hence the name. The place turned out to be, in Cassidy's words, "a real bonanza." The team gleefully harvested meteorites, giving each one a specimen name. One of them looked strange. Following their protocol, it was given the specimen number EETA79001. EETA referred to its location on Elephant Nunatak, 79 referred to the year it was collected, and 001 denoted that it was the first of the meteorites picked up at that location.

As in every other field season since the beginning of the project, the meteorites, including EETA79001, were brought to the Johnson Space Center in Houston. Researchers around the world sent applications to work on each meteorite and, following Cassidy's convention, successful teams were sent small pieces that they could work on in the lab, exploring their mineral structure as well as the atoms inside.

One team noticed that EETA79001 had small pieces of glass inside. This kind of structure is only encountered in a few places where the temperatures and pressures get high enough to melt the rock. The melted rock forms a glass when it cools. One location where this deformation happens is a nuclear test site. The others are places that were hit by large asteroids. EETA79001 came from a rock that was produced by an asteroid collision. But what did that asteroid hit?

Another group took their sample and analyzed the chemical constituents of the glass particles inside the meteorite. They pulled out each tiny piece of glass and then remelted it. As they did this, dissolved gas was released. When they looked at the chemical structure of the gases, they found oxygen and nitrogen. They also found atoms of xenon, krypton, and argon.

This batch of atoms would just be a laundry list with no special meaning if it weren't for another scientific expedition that was happening in the mid-1970s. In August and September of 1975, NASA launched two probes, Viking I and Viking II, to be the first American missions to land on Mars. Both successfully landed, leaving an orbiter around Mars to relay signals back home. The probes had chemical laboratories on board to make the first analyses of the

Martian atmosphere. When the signals came back to Earth, the Martian atmosphere was found to have oxygen and nitrogen, as well as atoms of xenon, krypton, and argon.

The meteorite researchers knew of these discoveries when they assessed EETA79001. When they compared the gases released from the glass inside the meteorite to the atmosphere of Mars, they found the gases were identical. EETA79001 was a chunk of Mars, blasted off by an asteroid impact, only to fly into space and enter Earth's atmosphere. Even the chemistry of the Martian atmosphere was preserved inside its atoms.

What about the meteorites that landed in Nigeria, France, and India? They, too, were pieces of Mars that were shattered off the planet and landed on Earth. So was the rock that was said to have killed the dog in Egypt in 1912. Bill Cassidy once wrote a lament of the misfortune of that dog. Think of the odds of an asteroid hitting Mars, spalling off a rock that gets caught by the Earth's gravitational pull and lands on Earth in the exact spot the hound was walking in Egypt at nine a.m. on June 28, 1911. Very, very, small.

If the El Nakhla dog's story was a case of severe bad luck, then that of Rebecca Score qualifies as good fortune. Score, lab manager of the Johnson Space Center's meteorite collection, joined the ANSMET team when they went hunting in the Allan Hills region, 200 miles from McMurdo Station in 1984. She picked up a dark meteorite and duly labeled it ALH84001, the first meteorite of the season. Little did she know the controversy her find would kindle.

The meteorite sat in the collection, garnering the sporadically intense scientific interest that comes with being a Martian meteorite. Teams applied to work on it, and samples were sent around the world. The team at the Johnson Space Center got to work and noticed something strange under the electron microscope. Small rods, less than a tenth of a millimeter long, lay jackstraw across the rock. Though the rods were tiny, they had a shape that the investigators had seen before. They looked like bacteria alive today and in the fossil record.

Bacteria in Martian rocks? The Johnson Space Center team decided to search for markers of life in the rocks, and with this new goal in mind, they set off to analyze the chemical structure of the rods. They found small shards of magnetite and aromatic compounds that are typically associated with certain bacteria. The scientists' conclusion was as bold as it was controversial. ALH84001, a Martian meteorite from Antarctica, contained evidence of fossil life on Mars. The announcement of the find was made at a press conference by President Bill Clinton.

To nobody's surprise, the announcement set the scientific community on fire. Critics got to work to show that the extraordinarily small size of the so-called bacteria would be unique among all known life-forms. And because other rocks are known to have magnetite and similar aromatic compounds, the evidence was not convincing. While there is no definitive evidence of life in ALH84001, that doesn't mean the interpretation is wrong. We simply do not have enough data to contradict the original hypothesis that fossil life is present. As Everett Gibson from the Johnson Space Center said in 2006, on the twentieth anniversary of the announcement, "Despite whether you are a believer of the hypothesis or not, it has clearly been the guiding idea for the development of the new interdisciplinary field of astrobiology." In our age of Mars rovers, data from space will enable us to analyze Martian meteorites just as the Viking landers did in the 1980s.

After chatting with Johnny in 2018, we desperately wanted to find pieces of Mars, let alone any chunk of rock from outer space. Our hunt for fish was on the rocks, not the blue ice. But we would need to traverse the ice on our snowmobile runs to different mountains. In one of our initial traverses I stopped the snowmobile to scan the surface for meteorites. I found an ocean of potential ones.

A sea of rocks, each with a black rind, were strewn across the surface. Most of them had pores. Thinking I'd hit the jackpot after only a few moments on location, I collected a sack of these rocks to return to the meteorite hunters. But the mountains in our area were rich in basaltic rocks. Basaltic rocks are ancient lavas that cooled, forming pores and irregular surfaces. The winds in this area are so intense that the outer surface of exposed rocks is sculpted by grains of dust and snow, essentially sandblasted. I had collected a bag of basalt, not meteorites. Antarctic science is a world of humility.

## Dusting Off the Poles

Our planet is a dusty place. Each day tiny particles from space enter the atmosphere and accumulate on the land, sea, and ice. By some estimates, as much as 60,000 tons of debris from the solar system rain down each year. That large mass dwarfs that of meteorite falls—a mere 50 tons of meteorites might land on Earth in the same interval of time. Some particles are barely larger than a molecule. Others, such as micrometeorites, might be about half the size of the period that ends this sentence. If there is an undisturbed flat surface, such as the roof of your house, a three-foot square patch might be hit by one of these each year. Smaller particles likely arrive in much greater numbers.

This dust is rich in history: it consists of the remnants of stars, or the primordial material that made our sun and solar system, or even parts of other planets that entered space after asteroid collisions. Scientists have launched probes into space to collect stardust, looked at sediments at the bottom of the ocean, and set up giant vacuums around the planet. One Norwegian amateur astronomer surprised the scientific community about a decade ago, devising methods that anyone with simple tools, such as a magnet, can use to collect micrometeorites and dust from their roofs, even in large cities. In most cases, though, it is challenging to separate the signal from the noise with these collections. Our world is full of dust of all kinds, from exhaust to the debris of human activity, and finding places where the cosmic dust is in pristine shape can be a challenge. But there's one obvious candidate: Antarctica. Its remoteness from all ordinary human activities, and its continent-sized sheet of pure ice, has made it a natural laboratory for scientists to collect and study stardust.

In 2006, three German scientists set off to mine for dust in the ice of East Antarctica, near Germany's Kohnen Station. This tiny base is open for a few months of the austral summer and houses only twenty people. It sits on the ice about 150 miles from the coast. At high altitude, and removed from human activities, the ice there is nearly pristine ice.

The team selected a site upwind from the base to guarantee sample purity, then carefully dug out 1,000 pounds of snow just below the surface. Using their knowledge of the ice, they were careful to collect the snowfall from a layer that represented only a single year. The team then shipped the sample back to the lab in Munich while keeping the compacted ice frozen for the entire trip home.

Meteorite and dust laboratories have clean rooms, where researchers suit up in gowns and slippers and the air is filtered to remove contaminants. In the clean room in the Munich lab, they melted the ice and, using fine filters, collected all the solids. Then they probed this dust with mass spectrometers at another lab in Germany to separate and analyze the individual atoms contained inside.

The team was surprised to find a strange element in the ice: iron 60. This version of iron is produced in only two known places. It is the product of nuclear reactions such as those that occur in reactors.

The scientists knew, given the prevailing winds, and the great distance between any nuclear reactors and the Kohnen Station, that a nuclear plant was not the source. The only remaining possibility was that the iron 60 came from a large supernova in outer space.

How could the remnants of a supernova end up on Earth? The researchers proposed that it was due to the way in which our sun moves through the Milky Way galaxy. Just as the planets of the solar system orbit around the sun, the sun itself travels through the Milky Way. Currently, the solar system has entered a vast field of space known as the Local Interstellar Cloud. This field, about 300 light-years across, is a field of dust created by a star that exploded eons ago. The solar system moves about 50 miles per second through this field and, like a convertible car with the top down, the field's dust enters our orbit each day.

We have only begun to explore Antarctica for the most ancient visitors to our planet. The clear air and lack of pollution at the South Pole mean that there is no soot or particles from human activity or weathering rocks. Something else is in the air. To catch it in that purity, scientists at the South Pole Station set up a giant vacuum, with a filter at one end. They placed this collector, known affectionately as the Cosmic Dust Sucker, far from the station itself so that there would be no air pollution from exhaust. The filter collects thousands of micrometeorites each day, representing fragments from all over the solar system as it formed.

Some of the particles hold special importance. Their age suggests they were formed *before* our sun and solar system. They show, for example, how certain kinds of stars, known as red dwarfs, produce the carbon that composes our planet and our bodies.

The Cosmic Dust Sucker was set up in 2019 and quietly collects visitors from space every day in the summer. It's just beginning its work. Like satellite missions that travel for years to visit other planets, the Cosmic Dust Sucker quietly filters parts of space that have

The Cosmic Dust Sucker

traveled to us. The remoteness and extreme conditions of Antarctica, which created innumerable challenges for scientists and explorers such as Mawson and Bickerton, make it near ideal for studying other worlds.

Sitting nearby the Cosmic Dust Sucker, the South Pole Telescope is a marvel of engineering. Unlike a backyard telescope that visualizes the sky in visible light, this one uses a 30-foot aperture scope to analyze microwave light during daytime. Every piece of this machine follows a long supply chain all the way to McMurdo Station, then hundreds of miles more to the South Pole. And when things break, spare parts can be thousands of miles away.

Located at 9,300 feet above sea level on the polar plateau, the Amundsen–Scott South Pole Station is inhospitable, with average high summer temperatures of –15 degrees Fahrenheit. When a plane lands there after a three-hour journey from McMurdo Station, it leaves its engines running while passengers and cargo are quickly deplaned, and then immediately departs to make the return trip to McMurdo. Supply flights back up regularly because of bad weather or mechanical failures. The features that make the South Pole Station so challenging are also reasons it is one of the most special places on Earth to put a telescope. If cold, dry, and unobstructed by pollution or atmospheric disturbances, telescopes can have a high-resolution view of the sky. And since it is daylight or dark for six months at a time at South Pole Station, the instruments can monitor the sky full-time for part of the year.

The South Pole Telescope is designed to measure the oldest light in the universe, dating back to just 380,000 years after the Big Bang, when the universe was a hot, dense soup of charged particles. As the universe expanded and cooled, these particles combined to form neutral atoms, allowing light to travel freely for the first time. The cosmic microwave background is the afterglow of this event, a snapshot of the universe in its infancy. It appears nearly uniform in all directions, but scientists can detect tiny fluctuations in its temperature in small patches that allow us to unravel the mysteries of our cosmic origins. But these are very dim signals and require a special measuring device. Because of the clarity of the South Pole location, the telescope there can make measurements of the cosmic microwave background that are sharper and at a higher resolution than anywhere else on Earth. These discoveries reveal the hidden structure of the universe, one we cannot see with visible light. Differences in density of the signal can show the nature of mysterious dark energy and dark matter. Dark matter is completely invisible, yet by some estimates it makes up 80 percent of the entire mass of the universe. Its distribution in space needs to be measured indirectly, by analyzing the effects it has on nearby matter. Here is where the cosmic microwave background becomes useful. By measuring how the cosmic microwave background gets distorted or deflected at a very fine scale, astrophysicists can infer how patches of dark matter warp space in different parts of the sky.

Fragments of the solar system have rained down to Earth since the origin of the planet, and many have become trapped in polar ice. Like artifacts and shards from archaeological digs, meteorites are bits and pieces that can be used to reconstruct a larger history. Some Antarctic meteorites formed on small rocky bodies that orbited when the solar system was still a swirling disc of dust. The atoms inside those meteorites reveal the chemical makeup of the earliest celestial bodies and answer questions about why planets and asteroids contain different minerals. Other meteorites reflect different phases of the history of Mars and the moon. The most common lunar meteorites come from ancient lava flows on the moon. Work on the atoms inside the grains of these meteorites tells us when the lunar landscape was volcanically active and, later, when it became quiescent. Martian meteorites tell us about the early atmosphere and mineral composition of the planet. Mars was once a tectonically active world and meteorites give clues to the planet's transition to an inactive one.

While moving ice yields secrets about the history of the solar system, it also is essential for understanding Earth. The amount, distribution, and movement of ice on Earth is the result of the interaction between the oceans, land, and the atmosphere. The more we look, the more we find that ice becomes a keystone in understanding our planet's past, present, and future.

# CHAPTER 5

# A BRIEF HISTORY OF ICE

We were fighting sleet and wind after yet another fruitless day hunting for fossils on Canada's Ellesmere Island. It was July 2000, and our scientific target was the fish with arms and legs but, with our existing haul of only a handful of ancient bones and scales, we were still years from eventual success. More immediately, the three of us were a strenuous hike from camp. Our tents were set near the head of a narrow fiord, where a miles-long tongue of pack ice was engulfed by steep red sandstone cliffs. The return hike presented us with two options: either fight slippery boulders and scree slopes along the cliffs for the 5 miles back to camp, or follow raised beaches and watch out for polar bears on the adjacent pack ice. Knowing that walking miles over ice-covered rubble was a recipe for torn knee ligaments, we decided to take our chances with the bears.

Our beach traverse was interrupted by a prominent dark shape on the horizon. The sighting was unusual: with no trees, power lines, or buildings to mark perspective, it could have been a moderate-sized entity half a mile away or a huge one 4 miles distant.

Worried that it was a lone bear working the coast, or perhaps a wayward musk ox, we glassed it with our binoculars. The situation became stranger still. The object in our eyepieces was sitting stationary on the beach.

As we approached, the apparition came into clearer focus. It wasn't a single object; it was composed of collapsed wooden boards and slats weathered by the wind. We were now confronted with a mystery. Clearly somebody had taken great care to pile this debris here, but we were more than 300 miles from the nearest human settlement, an Inuit community of about 160 year-round residents. This presence became mysterious: with no roads, trails, or local landing strips, the only way in or out is by helicopter or on foot.

Looking closely, we saw that a few wooden strips were covered with print that was faded with time and exposure. The string of consonants in each word suggested the text was likely Scandinavian. The wood of the crates was windblown and worn gray, clearly having endured many Arctic winters of wind, snow, and extreme cold. The slats were not the only aspect that looked old. The nails that held the crates together had been pounded by hand, clearly made before factories could mass-produce them.

The weathering of the wood, the nails crafted before the Industrial Age, and Scandinavian text pointed to one possibility: we were standing near objects frozen in time like the fish fossils we were hoping to find. Once we understood what the broken and weathered boards represented, we stood in silent reverence: these were lost artifacts of one of the greatest episodes of exploration and scientific discovery in all of human history. And the story began more than a century before our hike on that beach.

In the mid-nineteenth century, Fridtjof Nansen trained as a scientist and oceanographer and wrote seminal treatises on the nervous systems of marine creatures for his dissertation. He was a man of immensely broad talents—a champion skier and skater in his youth, Nansen developed a passion for outdoor exploration. His interests coalesced in a profound attraction to explore northern polar regions. The object of his first quest was Greenland, which, in Nansen's time, had not yet been crossed east to west. In a departure from the standard procedure for polar expeditions, which consisted of large, heavily supplied crews, he took a small team of expert skiers from the eastern side of the island hundreds of miles across the ice cap to the western side. His success established him as a major player in polar research and lit a fire that was to consume him in the following decades. His achievement was also a confirmation of his approach to polar logistics. Nansen viewed other expeditions as bloated by big crews with unnecessary gear; his plans would be lean and mean, borrowing techniques from alpine populations and Inuit communities.

Nansen returned to Norway in 1889 with an entirely new quest in mind: to be the first person to travel to the North Pole. This desire was as much to achieve scientific breakthroughs as it was for the glory of conquest. At the time, the polar north remained a land of mystery and tragedy. Speculations, bordering on wild fantasies, about the region abounded. The extent to which land, ocean, and ice extended to the North Pole was in dispute. Hundreds of explorers had already been lost as their ships sank, they succumbed to scurvy, or they ran out of supplies during long winters in the ice. Nansen wanted to be the first person to see the extreme north, but he knew that conventional means wouldn't likely get him there.

Nansen was a student of polar logistics and found inspiration in unlikely places. In 1870, the U.S. Navy joined the international fervor to make a run for the North Pole. They retrofitted a British ship, renamed it the *Jeanette*, and established a crew led by a young prodigy, Lieutenant Commander George De Long. The *Jeanette* set off along a route that took it from the Bering Sea near Alaska west to Siberia. But after two years, the ship got stuck in pack ice. After the crew depleted most of its supplies, the ship was crushed by the floes. Some of the crew returned south intact. But De Long, along with twelve others, perished on the expedition in 1872.

Years after the death of De Long and his crew, the tragedy of the *Jeanette* provided constant reminders of the dangers of Arctic exploration. In 1884, Inuit reported finding broken boards, kitchenware, and other wreckage from the *Jeanette* along the coastline of Greenland. Somehow the debris drifted more than 1,000 miles in four years. To the polar exploration community, the artifacts of a lost expedition strewn across the Arctic were a cautionary tale. To Nansen, the debris formed the germ of a plan.

Nansen asked: What if the transport of the *Jeanette* wreckage reflected a steady current that ran from Siberia all the way to Greenland? What if he could design a sturdy ship that could resist being crushed by ice and could be carried by the current toward the North Pole? The idea was to let the current do most of the work to get him north, and then, as he hit ice, Nansen would make a short sledge run to the pole. He paid attention to even the smallest detail of his new project and studied Inuit ways of life for design inspiration for his sledges, cooking stoves, clothes, and other expeditionary gear.

Nansen set off to find funding for his idea. Presenting his plan to the Royal Geographic Society, then a home for most active and retired polar explorers, Nansen was met with a combination of skepticism, concern for his mental health, and outright derision. The famed American polar explorer Adolphus Greely, himself no stranger to polar tragedy, having led an expedition whose stranded survivors supposedly partook in cannibalism, proclaimed Nansen's idea an "illogical scheme of self-destruction."

But Nansen knew that an attempt for the North Pole could stir national pride. He concentrated his efforts at home and gave a rousing speech to Norway's parliament focusing on his country's special relationship to the pole, and the ways that its citizens were uniquely capable of attaining it. He ended up getting support from both the government and Norway's royal family to build a ship and hire a crew made up exclusively of Norwegians.

With funding in hand, Nansen hired one of the most renowned naval architects of the time to design a ship that was small, nimble, and sturdy, with a rounded bottom that would allow it to ride above the ice that would have crushed other ships. He named it the *Fram*, Norwegian for "forward."

In 1883, Nansen launched a three-year expedition to drift north in the *Fram* with the current, then sledge toward the pole. The *Fram* drifted thousands of miles, with the only problem being, as Nansen described it, "deathly boredom." When the ship reached the farthest north the current could take it, Nansen and a companion left the *Fram* to sledge the remaining way to the pole. While he didn't attain his goal, Nansen got farther north than had yet been recorded. And unlike any other expedition to date, Nansen's suffered not a single casualty or notable mishap. Nansen returned a hero and later switched careers to diplomacy, ultimately winning the Nobel Peace Prize in 1922 for his efforts rescuing refugees, including Armenian victims of genocide.

Nansen's voyage resulted in a six-volume treatise on the polar north, revealing that polar exploration is itself a science, with gear, logistics, and training reflecting studies of human physiology and methods developed by Indigenous Peoples. The *Fram* was to see two other expeditions. One was to take Roald Amundsen south for his successful attempt on the South Pole. The other took a small crew with a different Norwegian leader to Canada's Ellesmere Island to study polar geography, natural history, and climate in 1902. The smashed crates we encountered that July day in 2000 were from when the *Fram* explored our fiord during the 1902 expedition. We imagined that the crates had likely held supplies.

Fridtjof Nansen and the *Fram*

For me, the smallest discovery made by the *Fram*'s team was the most significant. As the *Fram* probed into what eventually became the Canadian Arctic islands, the youngest scientist on board, the ship's natural historian, found tiny fish scales inside the rocks of one of the fiords. These were from the extinct fish species that my colleagues and I were finding fossils of in the same fiord nearly a hundred years later. And that sledge that Nansen had designed using Inuit design concepts? I was to use a modern version of the Nansen sledge when I later ventured to Antarctica on my own expeditions on the polar plateau.

The shattered crates we saw on the fiord that July day in 2000 were a vivid symbol of scientific discovery in polar regions. Venturing the farthest north, farthest south, and then north again, the *Fram* catalyzed decades of discovery in the Arctic and Antarctic.

And the *Fram* expedition achieved something else: Nansen revealed a world of ice at the North Pole. To some, this revelation would have been the biggest surprise of all.

## Heading North to Paradise

The discovery of new trade routes, resources, and lands to claim were among the aspirations that led governments and rich benefactors to outfit ships for multiyear trips to the pole. For some explorers, however, the motivation included scientific discovery. These teams were driven north by a hypothesis about the inner workings of climate, oceans, and ice at the North Pole. This quest for knowledge even drove some to their deaths.

William Barents, born in 1550, was a Dutch navigator and cartographer who had a passion for finding an Arctic corridor free of ice connecting the Atlantic and Pacific Oceans. Barents envisioned a sea of clear water to the north of the pack ice of the Atlantic Ocean that he could use to shorten routes in the spice trade. Barents was of the opinion that twenty-four hours of continual daylight in the extreme north should make summer temperatures too warm for permanent ice, even at the North Pole. Further reducing the presence of ice, he surmised, was the saltiness of the sea, which chemically lowers the temperature at which ice forms. Together, these ideas led to the influential theory that an entire ocean extended, uninterrupted, from Alaska to Greenland and Siberia. This dream of an ice-free pole, known as the Open Polar Sea, drew explorers and their patrons to venture ever farther north.

The names of Barents, Baffin, Hudson, Frobisher, Scoresby, Ross, Melville, and Franklin—all explorers who tried to push to the farthest north or find an ice-free passage between oceans—are now scattered across the maps of bays, islands, and mountain ranges of the Arctic. And while the expeditions were outfitted by males, some of the most lasting contributions to polar science were made by a woman, Lady Jane Franklin.

In the eighteenth and nineteenth centuries, women lacked professional opportunities on expeditions. Lady Jane Franklin was born in England in 1781, and undeterred by these limitations, she became an accomplished traveler while her husband, John Franklin, led missions to Australia and eventually the Arctic. As a British naval officer and explorer, Sir John Franklin was the leader of what ultimately became known as the Lost Ship Expedition. In a well-known tragedy, his crew went to the Arctic in 1845, never to return. The fate of his ships, the *Erebus* and the *Terror*, was an enduring mystery for more than a century until remains of the vessels, along with artifacts of the expeditions, were discovered underwater in 2011.

Lady Jane Franklin

After her husband's disappearance, Lady Franklin led a decades-long quest to find him. She deployed her considerable influence to obtain patrons, ships, and crews for rescue expeditions in the hopes that her husband had found refuge in some far corner of the Arctic. Lady Franklin's passion and persistence catalyzed a new age of polar exploration. While the search for John Franklin was the stated goal, explorers used her patronage as an opportunity to push the motives that had driven them for centuries. For some, that goal included the hunt for the Open Polar Sea.

Influenced by Lady Franklin, American explorers and their wealthy benefactors got into the act in the years prior to the Civil War. Chief among them was Elisha Kent Kane. Born in 1820, Kane trained as a geologist at college in Virginia, but after falling ill with the heart condition of endocarditis he returned home to Philadelphia to find a less physically demanding position. A medical career fit the bill, and after Kane graduated from the University of Pennsylvania Medical School, he worked at a Philadelphia almshouse as resident physician. But Kane's lust for life in the field couldn't be restrained. Ignoring his heart problems, he entered the army as an assistant surgeon. The gig led him to gain expeditionary experience in China, Mexico, and the Philippines. Kane's talents and zeal were noticed in high places and he was appointed senior medical officer on the Grinnell Arctic Expedition of 1850–1851 in search of Franklin's Lost Ship Expedition.

Kane was part of the landing party that discovered the graves of three crew members of the Franklin expedition on Beachey Island. This success gained him renown and catalyzed his passion even further. Being a natural communicator, Kane wrote a book on his expedition and led a speaking tour on his Arctic experiences. He used his growing public stage to pitch a second expedition to the wealthy merchant Henry Grinnell. Ultimately obtaining Grinnell's backing, Kane prepared a ship and a crew, ostensibly to search for

Franklin along the coast of Greenland and Ellesmere Island (then known as Grinnell Land). But Kane's real goal was to find the fabled Open Polar Sea and to reap the huge payoff of new trade routes between the Atlantic and Pacific Oceans that would be created.

Kane set off in 1853, making stops in southern Greenland to round out his crew with expert hunters, sledge drivers, and Inuit interpreters. These personnel choices turned out to be lifesaving decisions on Kane's part. During the stops, however, Kane learned he had competition. A British steamer had set off to the same part of the Arctic at roughly the same time as Kane. Wasting no time in port, Kane got to the coast of northern Greenland as soon as he could. There, at a latitude of 78 degrees north, Kane's ship was beset by sea ice. Ultimately, Kane and his crew were trapped in a desolate Arctic bay for twenty-one months. Their timing was supremely unfortunate, as the crew experienced some of the coldest winters ever recorded in that region. Temperatures dipped to –70 degrees Fahrenheit and sources of fresh meat diminished as animals to hunt grew scarce. Kane's medical background, his resourcefulness, and the personnel he'd brought from southern Greenland saved the crew. He saw the value of eating raw meat as a means of staving off scurvy, and regularly doled out the little meat they could find. Using his interpreter, Kane bartered and collaborated with Inuit for supplies and food and developed close relationships with local populations. He lined the ceilings, walls, and floor of the hull of his ship with moss, tundra, and furs in Inuit fashion. When the temperature outside fell to –40 degrees in the dark of the Arctic winter, the inside of Kane's hull was a relatively cozy 40 degrees.

Despite his best efforts, the health of his crew started to fail. Feeling penned in by a ship frozen in ice, and seeing no chance of survival on board, nine crew mutinied. Putting their fate in their own hands, they departed to take their chances traversing south.

Woodcut of Elisha Kent Kane (center) in a heroic pose on his ship. A cozy scene amid a harsh Arctic winter.

Kane let them go amicably. And in an act as much about survival as it was about magnanimity, Kane ultimately welcomed them back after their effort failed.

Kane remained undiminished in his quest to find the Open Polar Sea despite these hardships. While his ship was beset, Kane sent sledges off in different directions to explore the natural history of the area and find evidence of his polar Eden. One sledge journey lasted several weeks and endured temperatures as low as –57 degrees. Making it to 80 degrees latitude, then the farthest north of any European or North American expedition, crew member William Morton stood on the summit of a cliff along the coast of northwest Greenland and peered into the distance. He famously described the vista in front of him: "Not a speck of ice. As far as I could discern, the sea was open, a swell coming in from the northward. . . . The surf broke in on the rocks in regular breakers."

That moment made Kane's effort worthwhile, despite the

hardships. The expedition didn't find evidence of Franklin, but it attained both farthest north and the first visual evidence of the fabled Open Polar Sea.

Kane had little time to celebrate. Worried that supplies, morale, and the physical condition of his crew would make surviving another winter extremely unlikely, he decided that his best option was to abandon ship and head south. After eighty-three days traversing the western coast of Greenland on foot, Kane and his crew found rescue as a sailing vessel passed by their position.

Kane returned to the United States a hero. Unfortunately, when the Civil War followed a few years later and stories other than those of exploration and discovery led the news and gained national interest, Kane slid to the margins of American history. Imagine if Neil Armstrong, Buzz Aldrin, and Michael Collins had returned from the moon only to be immediately forgotten. Two years after his rescue on the ice, Kane's endocarditis caught up with him. He died at age thirty-seven while returning from Britain after delivering his expeditionary report to Lady Franklin by hand.

One of Kane's enduring successes came from inspiring another medical student to probe the far north. Born in 1832, Isaac Israel Hayes, an MD from Kane's alma mater of the University of Pennsylvania, had been a junior member of Kane's expedition. Hayes learned Inuit techniques, acquired the polar exploration bug, and set off north in search of Franklin and the Open Polar Sea in 1860, three years after his mentor's death.

Hayes followed Kane's expeditionary tracks through northwest Greenland. His sledge excursions were modeled on Kane's, even to the point of trying to gain farthest north and get a glimpse of the Open Polar Sea. But Hayes's claims were suspect. His inability to use a sextant properly led to criticisms that he continually misconstrued his position. Moreover, unlike Kane's team, who saw open water in the region at the northern extent of their sledge runs, Hayes encountered floating ice blocks pushing against the Greenland coast. To him, the size and shape of the ice fragments suggested that an Open Polar Sea lay beyond. The ocean ice strangely didn't falsify the theory. Instead, to Hayes it became supporting evidence. In his reasoning, a southern ring of ice must encircle the open water.

Hayes and his crew explored the natural history of the region in addition to striving for the glitzier quest to find the Open Polar Sea. The coast of Ellesmere Island is home to cliffs with flat rocky layers that lie on top of one another like a birthday cake. Continuous rock layers such as these typically form by the deposition of sediments of ancient seas. Hayes mapped the structure of these rocks and, while analyzing them for his maps, collected small shelly fossils. While much of his collection ultimately was lost before it could be fully incorporated into the Smithsonian Institution, the shells told of what the polar region had looked like in the distant past.

Scientific advances in the polar regions don't proceed smoothly from one discovery to the next; they often take a circuitous course. After her husband's disappearance, Lady Franklin led a decades-long quest to find him. The search for an Open Polar Sea propelled Kane and Hayes to Ellesmere Island, where the latter found fossil shells inside the rocks. Nansen's *Fram* expedition decades after Hayes's return put an end to any notion of an Open Polar Sea and revealed that the North Pole sits on a sheet of ice covering ocean below. After Nansen, marine seashells in Arctic rocks made sense in a way Hayes never could have imagined. Hayes's claim that an ice-covered ocean must reflect the presence of an Open Polar Sea was confirmation bias bordering on wild fantasy. But the fossil shells he discovered *were* evidence of an Open Polar Sea—just not the kind he imagined.

## Science to the Bitter End

Humanity's encounters with Antarctica occurred millennia after those in the Arctic. According to an apocryphal tale, the first person to set sights on the Antarctic Circle was the seventh-century Polynesian navigator Ui-te-Rangiora. He purportedly named the area Tai-uka-a-pia ("sea foaming like arrowroot") because of the similarity of floating ice floes to his native arrowroot powder. Whalers, navigators, and sealers had encounters with islands and seas within the Antarctic Circle over the course of several centuries, and then a number of parties finally converged on the continent itself around the same time. The first recorded sighting of the continent was by a Russian expedition led by Fabian Gottlieb von Bellingshausen in 1820. Within ten months of Bellingshausen, British and American ships also sighted the continent, and claims of first discovery have been controversial ever since.

The lore of Antarctic exploration tells of a fervid competition between Norwegian and British teams to attain the South Pole in 1912. The classic story contrasts a lean, expert, and successful Norwegian team with a poorly prepared and bloated British one that led to tragedy. But there is a different tale hidden in the journals of these expeditions that reveals the extraordinary sacrifices people made for the sake of scientific discovery.

The British explorers Robert Falcon Scott, born in 1868, and Ernest Shackleton, born in 1874, led teams attempting to reach the South Pole in the years before 1911. Each team failed to reach their goal, but they each set successive marks for reaching farthest south. In 1909, Scott outfitted a ship, the *Terra Nova*, and publicly announced a plan to attempt the pole. Unbeknownst to him, Norwegian Roald Amundsen, born in 1873, had quietly prepared a run for the South Pole by deploying techniques adapted from Nansen,

Kane, and Inuit Peoples. Amundsen kept his plan secret, even from his patrons and crew, on his voyage south. Scott received word of Amundsen's expedition while he was in the final phases of outfitting his ship in Australia.

The challenge of reaching the South Pole is that it is not only extraordinarily cold, but that it lies at an altitude 9,200 feet above sea level. Moreover, ships cannot sail close to the pole because of the permanent ice that lies between the ocean and the glaciers that ascend thousands of feet to the polar plateau. The entire trip is perilous at any altitude because of near constant wind and ice that is roiling with hummocks and crevasses for hundreds of miles.

In the summer of 1910, Amundsen set off for Antarctica in the *Fram* within months of Scott launching in the *Terra Nova*. The composition of the Norwegian and British teams could not have been more different. Amundsen brought nineteen crew, consisting of fast skiers, expert sledge drivers, and skilled navigators, on a streamlined expedition of the kind that Nansen had used in his attempt for farthest north. Amundsen selected his base close to the polar plateau with a single goal: being first to the pole. Scott, meanwhile, sailed with sixty-five crew, twelve of whom were geologists, zoologists, meteorologists, and photographers, on an expedition that attempted not only to gain the pole but to do science and make a record of conditions there. While he prepared for his run to the pole, Scott's crew sledged around his base off of the Ross Ice Shelf to study and collect samples at penguin rookeries. They also headed inland to explore geologic features, and over ice and the ocean to make measurements of the seas and weather patterns. Scott's teams were the first to encounter fossil fish scales in rocks, the surprisingly arid climates of the Dry Valleys, the red liquid oozing out of Blood Falls, and other discoveries that have launched scientific studies ever since. Amundsen's team, on the other hand, was deploying their expertise in the science of polar exploration to launch a lightning run to the South Pole and back.

Both Scott and Amundsen had selected the members of their teams for the final run for the pole with traits that mattered to them. Scott chose the most durable members of his crew, including Edward Wilson, his dearest friend on the expedition. Wilson, the senior scientist on the *Terra Nova*, was a polymath—he practiced as a physician and was a gifted artist, natural historian, and keen observer of both the natural world and human nature. Scott's crew nicknamed him Uncle Bill for the way he tended to their physical and emotional needs with his unflappable nature. Wilson never missed an opportunity to make scientific observations, recordings, or drawings. Amundsen's polar team included a famed champion skier, navigator, and sledge driver, along with his own closest colleague from expeditions to the Arctic.

On November 1, 1911, Scott's team of five men set off on the trek. Amundsen had launched his attempt with a sledging team of four Norwegians two weeks earlier. When Scott's team ultimately arrived at the South Pole on January 17, 1912, they were despondent to find a tent with a Norwegian flag extending from it, as well as a note from Amundsen to Scott. Unbeknownst to Scott, the Norwegians had arrived at the pole over a month before and were almost back at the *Fram* by the time that Scott arrived. Scott's crew depleted much of their energy on the sledging and ascent, and a photo of them at the time records the scene. Scott, in the center, has pure desperation and a hint of defeat on his face. Others look withered and exhausted, except for Wilson. He stands at the left with his hands on his hips facing the camera with a determined glare. After snapping this photo, Wilson mustered the energy to sketch, document, and analyze the tent and cairn left by the Norwegians in his field notes.

Now physically spent and emotionally withered, Scott and his

Robert Falcon Scott (standing, second from right) and his crew at the South Pole. Their faces tell the tale—all beaten by exhaustion and the knowledge that Roald Amundsen arrived earlier. Except for Edward Wilson (undeterred, at left). Wilson went on to make scientific observations.

companions had an arduous and dangerous trip home. The return from the pole meant descending the Beardmore Glacier, a wall of ice that drops 7,200 feet into a valley lined by rocky cliffs. By the time the five explorers arrived at the glacier in early February, they were exhausted from the haul, despondent about their loss of the polar quest, and at the brink of physical failure. During this descent, one of the most robust team members, Edgar Evans, took a fall on the Beardmore Glacier and was becoming dangerously weaker with every passing day.

In the face of hardship, desperation, and danger, Scott's team stopped off at the Beardmore Glacier—*to study rocks.*

Scott wrote in his journal for February 9:

> We lunched at 2 well down towards Mt. Buckley the wind half a gale and everybody very cold and cheerless.

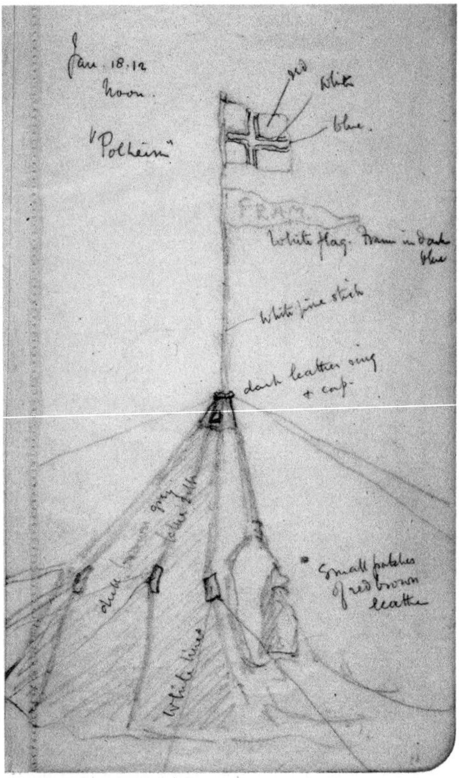


Ever the scientist, Edward Wilson overcame fatigue and disappointment to make scientific sketches of the tent and flag that the Norwegian team had left at their South Pole camp (labeled "Polheim" in the diagram) weeks earlier.

> However, better things were to follow. . . . I decided to camp and spend the rest of the day geologising. . . . From the last Wilson, with his sharp eyes, has picked several pieces of coal with beautifully traced leaves in layers.

Wilson found fossil leaves in rocks a little over 100 miles from the South Pole.

Wilson wasn't finished making scientific discoveries. The next day, February 10, he wrote:

> Some of the weathered blocks which we found on the moraine were of good burning shiny crystalline

> coal. . . . But the best leaf impressions and the most obvious were in the rotten lumps of weathered coal which split up easily to sheath knife and hammer. Every layer of these gave abundant vegetable remains. . . . Most of the bigger leaves were like beech leaves in shape and venation.

Wilson's fossilized beech leaves were from a different geological layer than the one he'd found the previous day. Wilson found two different kinds of fossil plants from two different geological eras on the descent of the Beardmore Glacier.

One week after these fossil discoveries, Edgar Evans became unable to move. Delirious, he then succumbed to the elements. In the following weeks Lawrence "Titus" Oates, another hardy member of the team, grew frail from frostbite in both feet and, fearing he was weighing down the group, walked out of the tent and into a ferocious gale to die with the epic final words: "I'm just going outside and may be some time."

With rations and fuel depleting, the remaining three members of the expedition—Scott, Wilson, and Henry Robertson "Birdie" Bowers—were slowed on their trek to a supply depot by persistent gales and ever weakening bodies. Bad weather trapped them in their tent only 11 miles from resupply and 60 miles from a sledge team sent as support. With their energy sapped by hauling sledges for 600 miles, then up and down almost 10,000 feet of elevation in extreme cold, the three ran out of food, fuel, and time.

A team arriving seven months later found Scott's tent, sledges, and cargo. When they entered the tent, which was in pristine condition, they found three bodies. Scott lay partially out of his sleeping bag with his left arm extended over Wilson, his longtime colleague. Outside the tent were Scott's sledges. One of them contained nearly 80 pounds of fossil plants Wilson had collected six

The fossil plant *Glossopteris*

weeks before their deaths. Even with their energy spent and bodies giving out, Scott's team never jettisoned their scientific cargo.

Their precious haul was a collection of fossils of a plant known as *Glossopteris*, a species that lived about 300 million years ago. While southern beech, fossilized in lumps of coal, was not part of the collection on the sledge, its presence was recorded in Wilson's diary. Similar fossils discovered since that expedition have been dated to be 5 to 7 million years old. The key point is this: *Glossopteris* is a tropical plant, and the southern beech is a temperate one. Neither species was ever thought to be polar. Wilson's ancient plants lie about 100 miles from the South Pole—a place devoid of plant life today. These discoveries invite the question: If the continent of Antarctica had been, like the Arctic, tropical or temperate at different times in Earth's history, when did polar regions freeze?

## Fevers, Chills, and Open Polar Seas

The movement of glaciers leaves traces inside rocks and on their surfaces. As massive bodies of ice, glaciers ebb and flow with changes in temperature. When they shrink and grow, they gouge bedrock. These gouges can extend more than the length of a football field. Some of the most famous glacial marks lie inside Central Park in Manhattan. Sheep Meadow, just north of the famous Central Park Carousel, has dark schists whose deeply carved surfaces are visible today; these grooves were made as the glaciers retreated from Manhattan 12,000 years ago.

Other evidence of glacial movement comes from the action of melting ice. As glaciers melt, they generate torrents of rushing water. These powerful rivers carry everything in the landscape in their wake, from large boulders to smaller pebbles. After the water recedes, this debris consolidates to form geological layers with a distinctive glacial signature. The sediments produced by melting glaciers are a jumble: pebbles lie next to larger stones, all in no particular order. This detritus stands out in a cliff and serves as a geological signpost of a glacial age. The atoms inside the rocks, being influenced by freezing and melting ice, are also transformed. Glacial meltwater has a different chemistry from liquid water and leaves distinctive chemical traces in the rocks as well.

With scrapes on rocks, jumbled sediment layers, and chemical patterns of meltwater in mind, we can look to the rocks of the world to find evidence of when and where massive amounts of ice were once present during Earth's history.

Rocks currently in Antarctica and the Arctic islands have been drifting about the globe for millions of years. The tectonic plates that cover the Earth's surface have been moving over eons of time, so much so that plates that were once close to the equator now sit at the poles. So, there are two reasons why rocks at the poles today may have lacked ice in the past. The first possibility is that the Earth was warmer then and ice couldn't form. The second one is that the rocks currently at the poles drifted from warmer equatorial environments. Accordingly, tracing the history of polar ice means assembling the map of the world through time.

Cartographers of the distant past deploy an arsenal of tools to estimate the age and geographic coordinates of rock formations. For age, the best methods involve the "atomic clocks" inside mineral grains. Certain atoms decay into others at known rates, and these rates can be used to estimate the ages of rocks and geological layers with varying degrees of precision. Plant and animal fossils can also play a role in determining the age of rocks—fossils form a succession of layers that can be used to determine their relative ages. Once we know the age of a particular geological unit, then we need to know the location where it formed.

Features inside rocks can also serve as a kind of low-resolution GPS to tell the latitude and longitude of their origin. Iron-rich rocks are particularly useful. The iron-rich minerals in ancient rocks can act like tiny magnets. To determine ancient latitudes from these samples, geologists measure the tilt of the iron fragments in the rock. The steeper the tilt, the closer the sample formed to the ancient magnetic pole, and thus, the higher the latitude. To determine ancient longitudes, scientists measure the angle of the iron fragments to geographic north. Because the location of the magnetic north pole can migrate over time, scientists need to make corrections to orient the rocks to where they lay in the distant past.

More geographic information comes from the structure of the rocks and geological layers themselves. Rocks formed in oceans have a different geological structure from those formed on land. Moreover, geologists can look for similar rock layers or fossils in different locations to assess whether geological formations that are far apart today were once connected in the distant past. For example, the fossil plant *Glossopteris*, discovered by Robert Falcon Scott's team on the Beardmore Glacier of Antarctica, is similar to specimens found in rock layers in Africa, Australia, and South America. This correlation is one line of evidence that supports the idea that these continents were connected in the distant past. Using these kinds of clues from different layers of rock from around the world, the positions of land, water, and ice through time can be pieced together like a planetary jigsaw puzzle.

For Earth's first 2 billion years, oceans opened and closed as continents joined together only to separate again. Mountains and sea levels rose and fell. While Earth's most ancient rocks hold evidence of these geological upheavals, they do not show two important things: there are no debris layers and there are no striations on rock. Large ice sheets did not exist during the first 2 billion years of Earth's 4.56-billion-year history. While ice might have sat on isolated mountaintops, the poles, along with the rest of the planet, were virtually ice-free.

The most ancient ice sheets were discovered almost by accident. Arthur Philemon Coleman was born in 1852 to a family of Methodist ministers in Quebec. Coleman's parents cultivated his interests in art, science, and philosophy to the point where each became a focus throughout his adult life. He trained as a geologist in an age before specialization and was as interested in minerals and fossils as he was in history and art. Coleman was attracted to fieldwork, and he spent months in the Canadian Rockies studying rock layers while painting watercolors of landscapes to give to friends. Renowned for his vigor, Coleman scaled numerous peaks throughout his life. During one trip to some mountains in South America, he hauled a tray up a mountaintop over 10,000 feet high. His colleagues were aghast when he used the tray as a sled to slide down the snowy incline at a blistering speed. What was remarkable wasn't the speed or the risk he took—it was that Coleman did this feat well into his eighth decade of life.

Coleman was a keen observer of rocks, reading their layers like pages of a book to tell the story of the history of the planet. During one trek to the rocks near Lake Huron in Canada, he noticed some layers with the characteristic signature of glaciers—striations marked their surfaces and rubble debris lay inside. What's more, these were some of the oldest known rocks. He'd found evidence of the earliest glaciation on Earth. Geologists revisiting these cliffs decades later confirmed Coleman's glacial hypothesis and showed the layers were nearly 2.2 billion years old. Coleman had discovered the first time when ice covered large swaths of the Earth, including the poles, for a relatively short period. Keep in mind that in geological terms, "short" is about 50 million years.

After that icy interval, Coleman discovered, Earth returned to its baseline condition of the previous 2 billion years—it was a largely warm, occasionally hot place. Then, one iconoclast showed just how dramatically the climate of the planet can flip.

Colleagues who have worked with Paul F. Hoffman, now retired from Harvard University, describe the challenges of keeping up with him physically and intellectually. His brain and body are always going full-tilt. When he is not mapping rocks and thinking about ancient environments of the Earth, he is running marathons well into his seventies and often leaving fit twentysomething graduate students exhausted at the end of a day on the rocks.

In the mid-1990s, Hoffman was mapping the geology along the coast of Namibia in southwest Africa. Namibia is renowned for its vast mineral resources and for having some of the oldest rocks and animal fossils in the world. While studying near Namibia's Skeleton Coast, where raging surf and shipwrecks abut a desert landscape with roaming lions, elephants, and giraffes, Hoffman found something curious. One layer of rocks, 710 million years old, held a jumble of stones inside ranging in size from pebbles to small boulders. He knew what these layers meant: they were formed by glaciers.

Returning to the lab, Hoffman analyzed the chemical signature of the atoms in these rocks and confirmed they almost certainly were formed by the action of glaciers. A relatively mundane discovery became an explosive one when he looked at the magnetic structure of the iron inside the rocks. These geological compasses pointed to an unexpected conclusion: the rocks had formed at the equator, not at high latitude. Confident in the data, Hoffman and his colleagues drew the obvious conclusion: 710 million years ago, glaciers extended from the poles to the equator; land was covered by glaciers while the surface of the oceans was frozen solid. Our blue planet was once an entirely white one.

This surprising idea had been floating around in scientific circles for decades, albeit in esoteric places. Ever the geologist and explorer, Douglas Mawson had proposed a similar theory fifty years earlier than Hoffman. After being made Sir Douglas Mawson following his tragic Antarctic meteorite adventure, he speculated that glaciers covered Australia around 700 million years ago. In 1992, Joseph Kirschvink at Caltech developed the hypothesis of a frozen "Snowball Earth" and made specific predictions of what could be found in the geological record. The idea was overlooked because he published it as a seven-paragraph paper in a 1,348-page book. Working only few years later, Hoffman and his colleagues assembled the data to confirm the hypothesis conclusively.

Needless to say, the claim of an Earth frozen solid set off a raucous scientific debate. But layers with glacial detritus at the equator kept turning up in different sites across the globe. The hypothesis became inescapable. Scientists may disagree on the extent of the

Earth the glaciers covered, or how solid the glaciers were—some argue for more of a slushy surface rather than an icy one—but the conclusion was clear. There was an interval when the entire planet was as cold and icy as the poles are today.

Analyses of more rock layers led to more icy discoveries: evidence for glaciers at the equator cropped up in a slightly younger layer as well. This observation meant there was not just a single period when our planet turned into an ice ball, 710 million years ago; there was another one 50 million years later. And each frozen interval lasted about 10 million years. Each time the Snowball Earth thawed, the world immediately flipped to one with no ice at the poles or anywhere else. For these intervals, the world shifted from an icy planet to a hothouse and back again. Earth was suffering from fevers and chills.

Following this back-and-forth between hot and cold, the Earth entered another stable—and warm—state for hundreds of millions of years. From 600 million years ago to the present time, Antarctica has moved from just below the equator to the South Pole, while Greenland has shifted north from the southern polar region. Two relatively short periods of polar ice show how dramatically different the early Earth was from today's. About 460 million years ago and again 260 million years ago, periods of intense cold set in for about 20 million years. At these times the only ice was located at the South Pole, while the northern polar region lacked glaciers altogether. After these cool periods, the Earth returned to being a warm place. The ages when dinosaurs, mammals, and birds arose were warm times. In fact, some of these periods were downright hot. Around 55 million years ago, the world's temperatures rose to between 81 and 94 degrees Fahrenheit for 170,000 years. While that period was unusually hot (keep in mind that the average global mean temperature for the twentieth century was 57 degrees), it has become a case study on what happens to the planet during intense global warming.

Polar ice caps have only been a feature of our planet for roughly 10 percent of its existence. Our perception of what is geographically "normal" is skewed by the fact that the entire history of our species happened during a special time on planet Earth—one with ice at both poles. For most of Earth's history, despite relatively short and occasionally dramatic episodes of freezing, Earth had Open Polar Seas.

## Our Recent Freeze

Debris layers show that our modern world of ice started to form around 34 million years ago. The freezing was slow at first, covering high elevations and the South Pole. Over the ensuing 20 million years, the ice expanded beyond the pole to the rest of the continent of Antarctica. The eastern part of Antarctica was covered in ice by 13 million years ago, while the western part, which sits at a lower elevation, froze by 7 million years ago. With this deep freeze came a drop in the sea levels around the world and a biological transition of the entire Antarctic continent. Antarctica went from being home to tropical rain forests teeming with flowers, birds, mammals, and frogs 90 million years ago to a frozen continent impoverished of plants and animals. Nearly all life on the continent either went extinct or moved away.

Greenland, the Canadian Arctic, Alaska, and Siberia didn't get the memo. The northern polar region continued uninterrupted as a relatively temperate environment while the south was freezing over. The first evidence for glaciers in the Northern Hemisphere comes from 10 million years ago, when deep-sea cores reveal the presence of glacial activity. It wasn't until as recently as 3 million years ago that large ice sheets covered the Arctic Ocean over the North Pole.

With ice at both poles, the world entered a new phase: sea levels dropped, patterns of temperature and precipitation changed globally, species went extinct, and new ones arose. Our modern world was born.

While polar ice has come and gone, we're left with the question of what causes Earth's poles to freeze. Why did Antarctica transform from a tropical rain forest to a frozen world? Why did it freeze 30 million years before the Arctic? Answers lie in seeing the connections between rocks, oceans, and life.

## CHAPTER 6

# HOW TO FREEZE A POLE

Stacked lenticular clouds across the fjord were a sign that a polar windstorm was headed our way. Earlier in the season, the team nicknamed these formations "flying-saucer clouds" because they appear as disks piled one on top of the other, often above nearby peaks. Their beauty was part of their danger: the clouds' elongated shapes and wispy tendrils indicate high winds in the area. With the sky getting bigger and darker with each passing minute, our job was to get back to the safety of camp as quickly as possible.

That morning, Farish had sent a fellow graduate student, Steve Gatesy, and me on a trek to the coast in search of fossil bones. This was the second summer of the Greenland project, in 1989, and we already had found a few places that held a smattering of teeth and bone fragments from dinosaurs and other Triassic Period reptiles. Aerial photos and geological maps revealed vast rock exposures on the mountains by the sea, and we were hoping to find some new sites in the ledges, valley walls, and gullies carved inside them. Steve and I had spent the morning climbing a sandstone peak that terminated in a bluff overlooking the water and ice below.

Our hike was a bust from a paleontological standpoint—we didn't see a scrap of bone in the rocks. But a high point came from a rare quiet and windless lunch at the edge of a red cliff that had a sheer 1,500-foot-drop into the Arctic Ocean. Steve, a rock climber, was more than happy to peer over the edge at the icebergs, seals, and walruses below, while I was satisfied eating my lunch of granola bars and peanut butter at a safe distance. Surrounded as I was by red and green rocks sculpted by ice, wind, and water into a variety of shapes, I could have stayed there all day. But now we had to scramble half a mile down the mountain's elevation and then several miles more across the valley floor to return to base camp.

The descent took us down a natural staircase of alternating sandstones, siltstones, and shales where resistant layers formed flat ledges about every 50 feet. The rhythm of this traverse was occasionally interrupted by a hoodoo—a naturally eroding sandstone pillar that appears in one's peripheral vision like a ghostly lost hiker. Other than moments when we were jolted by rocky apparitions, we were moving so quickly, gazing only up at the sky and down at our feet, that we may have been passing promising fossil sites in our rush home.

Then we hit a red sandstone ledge that formed a platform about the length of two football fields. Hunching over to catch my breath, I looked down. Below me was an impression in the rock of a three-toed footprint, with impressions of fierce claws at the end. The print had the hallmarks of one made by a carnivorous dinosaur. I followed the direction of the footprint until I saw another, and yet another, and another still. The prints formed a trackway of about ten steps in series. But this trackway was crossed by a different set of three-toed dinosaur prints, which itself was crossed by still others. Dinosaurs had run all over this place 200 million years ago. I tried to keep pace with the dinosaurs' movements by placing my own feet in one set of tracks, but to move within its gait, I had to stretch my legs almost impossibly far. Looking up, I saw Steve's silhouette at the far end of the ledge also trying to walk inside the dinosaur tracks. A drone view of the scene would have shown two people on the side of a Greenlandic mountain by the sea ambling about and doing silly walks, a paleontological version of the classic Monty Python sketch. Steve waved urgently to call me to his find, and the two of us forgot about the weather for a few minutes and stood in wonder at what lay at our feet.

This ledge later became known as the Dinosaur Dance Floor for the hundreds of tracks of large and small dinosaurs that crisscrossed each other. Some 200 million years ago, this site sat at the margin of a large lake and dinosaurs etched the mud with their feet as they ran. But in our own time, with the winds building, Steve and I needed to stop walking in dinosaur tracks and pick up our pace to get back to the safety of camp.

Steve and I had seen prints like these before. The central toes were elongated and the two flanking ones were relatively shorter, giving them the appearance of being made by a large bird with big claws. In 1858, geologist Edward Hitchcock, then president of Amherst College, first discovered a set of these prints in central Massachusetts and he gave them the scientific name *Grallator cursorius*, *grallae* being the Latin word for "wading birds." Since Hitchcock's time, *Grallator* trackways have been found in Nova Scotia, Connecticut, and Pennsylvania and south along the eastern seaboard to Virginia and North Carolina. In places close to metropolitan areas, a single trackway might be the basis for a state park or a tourist attraction. By that logic, the number of tracks on the ledge in Greenland would be a dinosaur Disneyland. But the rocks in these different regions share more than dinosaur footprints. Versions of the same green and red sandstones extend from eastern North America to Greenland too. These types of rocks and footprints are also found in Morocco, Algeria, and parts of western Europe.

Two hundred million years ago, dinosaurs could have walked from what today is Greenland to Massachusetts and on to Morocco and Algeria without ever hitting ocean. The land was connected in a giant continent given the famously epic name—Supercontinent Pangea. The rocks in Greenland were deposited as the supercontinent drifted and as North America separated from Africa and Europe, forming the Atlantic Ocean.

Continental drift and the theory of plate tectonics are great unifying principles for understanding Earth and its history. These ideas explain more than the distribution of dinosaur fossils and red and green rocks. In fact, some of the things that moving continents can do are downright chilling.

While expeditions in the nineteenth century were searching for Franklin's Lost Ship Expedition, an Open Polar Sea, or national and personal glory, a lone French scientist was sitting at home sketching equations that would ultimately change the way we understand the polar regions themselves. Of course, that wasn't his day job. Jacques-Joseph Ébelmen, born in 1814, trained in math and mining engineering, ultimately taking a leadership position at the porcelain factory in Sèvres, France, one of Europe's most distinguished and enduring manufacturers of fine ceramics. He was renowned for his ability to devise new industrial methods, such as changing the process for making bone china or creating new casting techniques for building large ceramic objects. When he wasn't thinking about porcelain or ceramics, Ébelmen was developing chemical equations exploring his true passion—Earth's rocks and minerals. He devised new ways to synthesize minerals in the lab and theorized how they worked in nature. During these scientific

Steve Gatesy at a small ledge of the Dinosaur Dance Floor. Dinosaur footprints are in the foreground.

peregrinations, Ébelmen devised a theory that was so radical, so ahead of its time, and so outside the range of geological thinking of his day that his ideas were overlooked for more than a century.

Ébelmen was fascinated by the ways rocks erode and break down over time. He speculated that these weathering processes were central to the movement of the element carbon across the globe. If this idea seems obscure to you, you're not alone. It took the work of other scientists over the course of a century to understand how it explains much of the evolution of Earth and life. It even helps us answer the question of what causes Earth to get cold enough to form ice caps at the poles.

The most obvious (but incorrect) answer to that last question is a change in sunlight over time. Perhaps ice sheets emerged whenever the sun was getting dimmer over certain points in the Earth's history? Our sun produces solar radiation that brightens and heats the Earth by nuclear fusion reactions going on inside the star. These atomic processes are predictable because they follow physical laws of nuclear reactions that are constant over time. But there is a problem with thinking that a reduction in the sun's activity caused polar glaciers: the amount of solar radiation produced by a star like our sun only *increases* over time. Which means the Earth should be getting warmer with each passing year. If more and more solar radiation was hitting the Earth every year for 4.56 billion years, then why are there polar ice caps today but not in the past? If anything, we should be seeing an Open Polar Sea near the North Pole and a tropical paradise in Antarctica.

So if the amount of sunlight that hits the Earth is not responsible for its cold periods, or at least the recent ones, then what is? A key player is, obviously, the element carbon, and the different molecules carbon can be a part of, such as carbon dioxide and methane.

In the nineteenth century, French physicist and mathematician Joseph Fourier first proposed the ideas that ultimately became known as the greenhouse effect. Glass greenhouses stay warm in winter because they let light in but do not let radiated heat out; hence, the interior of the greenhouse stays warmer than the outside. An analogous process works in the atmosphere, but instead of glass trapping the heat, certain molecules in the atmosphere do the job. These molecules, many of which contain the element carbon, absorb the photons and energy that tries to escape the planet. Some of this energy leaves the atmosphere, but most of it gets deflected and remains on Earth. Without these gases—mostly carbon dioxide, methane, and water vapor—Earth's heat would escape into space and the average temperature of the planet would be a frigid –4 degrees Fahrenheit.

Svante Arrhenius, born in 1859, built on Fourier's theory and showed quantitatively how the production of carbon by industry could raise global temperatures. Since the Industrial Revolution, a major source of greenhouse gases in the atmosphere has been human activities that release carbon dioxide. The burning of fossil fuels and the production of gases that contain carbon, such as carbon dioxide and methane, coupled with deforestation, which removes plants that would otherwise extract carbon dioxide from the atmosphere, have caused an increase in temperature over the past century. But what was happening for the 4.56 billion years before humans were on the planet?

Answers to this question rely on Ébelmen's insight from the 1800s about carbon. But the scientific community's appreciation of this insight involved the work of another chemical genius who sowed and germinated the idea at a time when the science finally was ready to let it grow.

Harold Urey was born in 1893 in a town in rural Indiana to parents who ran a small farm. Both parents had college degrees and supplemented their farming with other occupations; his mother was a part-time teacher and his father was a lay preacher in a pacifist

Christian denomination. Harold's father died when Harold was six, and the insurance money was dedicated to young Harold's education. Lacking money for other things, the family moved about the country as Harold trained to be a teacher at different schools. His precocious mind, particularly when applied to the sciences, led other students to give him the prophetic nickname "Professor." With the start of World War I, Urey's skills as a physical chemist were in demand in factories that made armaments and explosives. At this phase of his life, Urey's scientific work placed him at odds with his pacifist upbringing. The situation grew even more challenging when, during World War II, Urey joined the Manhattan Project to design the first atomic bomb. He remained deeply conflicted throughout his life and in an interview in 1961 said, "I regret that scientific experiments—some of them mine—should have produced such a terrible weapon as the hydrogen bomb."

After World War II, Urey settled at the University of Chicago along with a number of physicists who'd followed their mentor Enrico Fermi, the first person to split the atom. Urey could now pursue more peaceful goals. He set off to apply his ideas of physics and chemistry to problems that had formerly been the sole domain of astronomy, geology, and paleontology. Working with a graduate student named Stanley Miller, Urey devised an experiment to show how the basic building blocks of life may have arisen on Earth 4 billion years ago. Urey won the Nobel Prize in Chemistry for discovering a new form of hydrogen. Ever humble, he skipped the award ceremony in Stockholm to be present at the birth of his daughter. In fact, one of his daughters became a prominent physicist herself, as did several of his grandchildren.

Urey had an elegant way of viewing planets and life, a view that stripped them down to their essentials—interacting atoms and chemicals. We see Saturn with its rings, Jupiter with its bands and red spot, and the Earth with oceans, life, and mountains. But to

Urey, living things and planetary bodies in the solar system can all be reduced to chemicals and atoms moving from one part of the system to another. In his eyes, understanding Earth and its history meant seeing how elements moved from oceans to the air into rocks and into the crust of the planet.

This quest led Urey unwittingly to the idea that Ébelmen had proposed more than a century before. Urey and Ébelmen surmised that, over long time scales, carbon in the atmosphere sits at a balance between the planetary events that produce it and others that remove it—sources and sinks. Imagine a world without humans, the one that existed for almost the entire 4.56 billion years of the planet's history. What was the main the source for carbon in the atmosphere? Today, the immediate source for much of the carbon in the air is human activity, such auto and aircraft emissions, manufacturing processes, and agricultural practices that release carbon dioxide and other greenhouse gases. But the equivalent of smokestacks and tailpipes for the billions of years before humans was volcanic activity. Volcanoes take liquids and gases from Earth's interior and eject them to the surface and into the atmosphere. The gases they emit contain carbon in different forms, from carbon dioxide to methane. We know from estimates derived from current activity that volcanoes can emit almost 300 million metric tons of carbon dioxide every year. While that number is dwarfed by emissions from human activity today, which produce as much as forty times more carbon, over millions of years of geological time volcanoes can pump out an enormous amount of greenhouse gases into the atmosphere. If this process were to go unchecked, the carbon in the air would accumulate at impossibly high levels and temperatures would become too hot to allow most of the kinds of life we see on Earth.

If volcanoes are the long-term source for carbon in the atmosphere, what is the sink that removes it?

The introduction of carbon into the atmosphere sets off a chain reaction of events. The carbon released by volcanoes in the atmosphere interacts with water vapor to form a very weak acid, known as carbonic acid. This acid cannot stay in the atmosphere long because, dissolved in water, it will rain down to the surface. When acid rain falls to Earth, it erodes rocks over time. Through the inexorable movement of water from land to the sea over millions of years, the carbon originally in the atmosphere travels with the eroded molecules from the rocks into the oceans. There, the carbon settles to the ocean floor and ends up in deep layers of the Earth. Eventually, the deep carbon is returned to the atmosphere by volcanic activity. Some recent estimates suggest that it takes 100 million years for one carbon atom to cycle from the interior of the Earth to the atmosphere to the oceans and back to the interior.

In terms of geologic time, Earth's climate sits in a balance between forces that increase carbon in the atmosphere, such as volcanism, and the formation of acid rain and chemical weathering of rocks that remove it. More volcanism means more carbon in the atmosphere and warmer climates. Likewise, any mechanism that serves to change how rocks chemically erode can reduce carbon in the atmosphere and reduce global temperatures. The breakdown of rocks does more than explain why we have canyons, valleys, and ocean cliffs—this weathering changes Earth's climate over long time scales. Anything that accelerates the chemical breakdown of rocks over long periods of time—changes in the amount of precipitation, the position of the Earth's continents, or the topography of the land—can cool the planet.

If interactions among volcanic emissions, acid rain, and rock weathering control climate, then there should be links between these geological events and global temperatures in the distant past. In 1983, a trio of scientists from Yale, the University of Florida, and Penn State tried to test that idea. Robert Berner, Antonio Lasaga, and Robert Garrels looked at the geological proxies for volcanic activity, acid rain, rock weathering, and other parts of the long-term carbon cycle in the rock record from 500 million years ago to the present. They measured variables such as the extent of exposed bedrock to estimate how much carbon was removed by rock weathering. The trio analyzed igneous rocks, former lava flows, to estimate volcanic activity during different eras in Earth's history. Chemical records and plant fossils were used as indicators of the temperature of the planet. Assembling these data in a dense paper that would become required reading for budding geologists of a certain age, they made a model that explored changes in the sources and sinks of carbon over time. The team found that, to a rough approximation, they could derive Earth's temperatures over large time scales by knowing the rates of the geological processes envisioned by Urey and Ébelmen. This was only a crude first attempt, but it was a glimmer that suggested that with more data, scientists could decipher the causes of long-term changes to Earth's climate system at key moments in the distant past, particularly moments when the Earth froze.

Ébelmen and Urey theorized that rocks, the atmosphere, and oceans were linked by the movement of carbon through them. It's hard to overestimate how ahead of its time Ébelmen's idea was in 1846: there was little understanding of the age of the Earth, nothing was known of how rocks weather, and his understanding of the carbon content of volcanic gases was guesswork. The world simply wasn't ready for his idea. It would have been as if somebody had sketched a design for a smartphone in the 1800s—lacking any context, the idea would languish. Ébelmen passed away at the age of thirty-seven, in the prime of his career.

To get to the bottom of complex criminal cases, we often hear that it's best to "follow the money." To understand the causes of warm and cold periods during the history of the Earth, the parallel is to "follow the carbon." Ébelmen's and Urey's theories about carbon movements related different parts of the Earth's land, oceans, and life to climate in unexpected ways. Understanding those linkages is the path one must take to see why our planet is capped by icy poles.

## A Cold Case

If you took a time machine to the early Earth, you'd find there are moments when you'd need a parka at the equator and others when a swimsuit would be useful at the poles. Over millions of years, our planet traverses the solar system while continents drift, oceans rise and fall, and volcanoes poke through the Earth's crust. Via this geological and astronomical activity, new kinds of living creatures arise and go extinct. Ice sheets develop when changes to the continents, oceans, and life align to make the world colder. These "perfect freezes" have shaped the planet and life on it, including our own existence.

As director of Columbia University's Lamont–Doherty Earth Observatory, Maureen Raymo leads a prominent scientific institution that has a history of exploring the seafloor, ocean currents, and plate tectonics. Given such an august and important post, one would think she has commanded respect throughout her career. Being a woman in a male-dominated field, however, has its downsides. Raymo has, by her own reckoning, been "treated as a secretary by older scientists on more occasions than I care to remember." A male member of her PhD thesis committee once called her work "a total crock." And later, a male department chair when she was on the faculty at MIT lectured her after a seminar: "I would caution you not to think you are smarter than you really are."

Maureen Raymo

Undaunted, Raymo has pursued a grand ambition throughout her career. She is, as she told a reporter in 2020, a "puzzler." Oceans, continents, and the atmosphere are pieces of a puzzle that tell us about what shaped the past and will form our future. And the movement of carbon through this system is a key to finding the answers.

While still a graduate student, Raymo learned that our planet's climate had cooled dramatically in the past 50 million years before the poles froze over. Her new puzzle to solve was to find the cause of that freezing.

A glance at a map of the continents of ancient Earth gave Raymo hints of a possible answer. The biggest change to the map came from the movements of the Indian subcontinent. Sixty million years ago, India sat below the equator as a separate island continent surrounded by water. By about 50 million years ago, it was moving north at the rate of 1 centimeter per year. Over millions of years, India crept ever northward until it slammed into Asia. Still pushing north, the crust at the border of India and Asia was the same density and had nowhere to go but up. Thus began the uplift that formed the Tibetan Plateau. Gradually, this slow uplift led to the origin of the highest mountain range, and the most extensive mountain plateau, on Earth. Today, the Tibetan Plateau continues to rise at about 1 centimeter per year. But erosion of the rocks happens so fast on the Himalayan mountains that their height rises somewhat less—as the mountains get pushed upward, their summits weather downward.

Raymo knew of this balance between uplift of the Tibetan Plateau and the weathering of rocks on its peaks. And, like many of her generation, she also studied the scientific paper by Berner, Lasaga, and Garrels on the long-term carbon cycle. Tying both ideas together, Raymo proposed that the rise of the Tibetan Plateau expanded the amount of rock to be weathered by acid rain. And since weathering causes a drop in carbon in the atmosphere, the entire Earth would have entered a period of slow cooling as the Tibetan Plateau rose.

Chemical analysis of ocean rocks from 60 million years ago to the present time appears to bear out Raymo's connection. Cores made into the deep ocean floor reveal that the atomic forms of strontium and lithium, which are produced by the weathering of rocks on land, become more common with the rise of the Himalayan mountains and plateau. The timing of this weathering coincides with the cooling of the planet.

These tectonic events heralded a period of ice on Earth. Starting 47 million years ago, the planet began cooling, but it was not until 15 million years ago that we first see major signs of ice in Antarctica. This fact is puzzling—the continent of Antarctica was sitting at the pole while the world was getting cooler. While the continent itself was cool, other events led it to be covered by ice miles thick.

Recall that 200 million years ago, all continents were joined as the single supercontinent, Pangea. The supercontinent initially broke up into two pieces: a northern landmass containing what is today Europe and North America and a southern one, Gondwana. The later breakup of Gondwana involved the separation of Antarctica from the other southern landmasses: Africa, South America, and India. As those continents separated 34 million years ago, they left Antarctica at the South Pole while they headed north. As they broke away, Antarctica became encircled by water. This water helped freeze the continent.

A new current was established by the ring of water around Antarctica. The circumpolar current, driven by continual westerly winds, flows around Antarctica clockwise if seen from the South Pole. No land obstructs the flow of water and wind as they circle the globe. With no barriers to slow or otherwise interfere with it, the current has become the most powerful on Earth. As any mariner of the southern seas knows, these are treacherous waters. But the circumpolar current involves more than water, wind, and danger—it influences heat. The formation of this flow essentially insulated Antarctica from the rest of the globe. Warm waters from more northerly oceans were no longer able to approach Antarctica. Consequently, the continent grew ever colder and colder, eventually freezing over entirely.

But as we saw from the rocks, a strange thing happened up at the North Pole when Antarctica froze. The North Pole and Greenland stayed relatively warm and ice-free.

A cooling planet alone does not tell us how both poles froze in the modern age. While carbon in the atmosphere can control the total amount of heat that the planet holds, other factors lie behind how heat moves across the planet. If global cooling set the stage for a cold Antarctica, and the formation of the circumpolar current was the trigger for the freezing of the continent, what happened to freeze the Arctic?

Water holds significant amounts of heat, and ocean currents

either insulate continents, as in the case of Antarctica, or move heat and moisture from one place on Earth to another. One of Earth's dominant ocean currents for the past 50 million years has been the Gulf Stream, which brings up warm water from the Caribbean and Gulf of Mexico into northern Europe and Greenland. Once at these cold northern latitudes, the waters lose heat and moisture to the atmosphere, becoming salty and dense. The density makes the waters sink into the deeper ocean and flow south into the Southern, Indian, and Pacific Oceans. After taking this path, the current returns to the Atlantic to begin the cycle again and head north.

A major mechanism for this current is the difference in salinity between the Atlantic and Pacific Oceans: the saltier the Atlantic is than the Pacific Ocean, the faster the Gulf Stream flows. Saltier waters are denser and descend more powerfully than less saline ones. This effect serves as an engine that drives the Gulf Stream.

Before 3 million years ago, the Atlantic and Pacific Oceans were connected, as South America and North America were separated from one another. Water flowed from the Pacific to the Atlantic via an open seaway between the continents. This current allowed the two oceans to exchange waters and remain similar in their salinity. Consequently, the Gulf Stream was weaker than it is today.

Starting 15 million years ago, two plates of Earth's crust started to collide into one another in Central America. The collision led volcanoes to form, and these volcanoes in turn formed islands between the continents of North and South America. The volcanoes grew into a giant arc that 3 million years ago formed the Isthmus of Panama. The continents of North and South America were now connected, and the seaway that once bridged the Atlantic and Pacific was blocked. The Atlantic and Pacific Oceans were now separated.

This new barrier set off a domino effect of changes to water, currents, and climate. With no seaway connecting the Atlantic and Pacific Oceans, the salinity of the two oceans began to diverge. Then the continual action of the Gulf Stream made the Atlantic become ever saltier than the Pacific Ocean. This change allowed the Gulf Stream current to become stronger. A powerful Gulf Stream brought ever more moist water to the frigid cold of the Arctic, and snowfall increased. Snow became compacted into ice, covering more and more of the Arctic Ocean and nearby islands each passing year. After several hundred thousand years of currents bringing moist air, then snow, and later ice to northern latitudes, the Arctic Ocean and nearby islands were covered by the polar ice cap.

With the Earth in a relatively cold state, small changes could now tip the planet's climate system into ice ages. One of the primary changes was astronomical. In 1941, Serbian mathematician Milutin Milanković hypothesized that regular changes in Earth's orbit can flip the planet from an ice age to a warmer one. As Earth orbits the sun, it interacts with the gravitational pull of other planets. Saturn and Jupiter exert the greatest effect on Earth's orbit because their enormous mass dwarfs that of other nearby planets. Milanković calculated that these interactions change the orbit of Earth in very predictable ways. The annual traverse around the sun shifts from being shaped like an ellipse to being more circular every 100,000 years. The axis of rotation of the planet swings every 41,000 years while the Earth wobbles like a top in a cycle that lasts 26,000 years. Milanković proposed that these changes conspire to affect the duration of the seasons and the amount of solar radiation that lands on Earth. With regular changes in the incoming and outgoing heat of the planet, there would be times when glaciers flow to lower latitudes, alternating with times when they retreat.

For decades these ideas sat as ingenious quantitative calculations lacking any real data. Then, in the 1960s, scientists looked at indicators of past climate changes in the seafloor and found that ice ages have come and gone over the past 3 million years with the regularity of a metronome that is largely calibrated to the orbital cycles Milanković predicted.

The stage was set—drifting continents, new ocean currents, and rising mountains caused both polar regions to freeze. Changes to Earth's orbit triggered the planet to change from glacial times to ones lacking in ice. The movement of ice changed everything.

## Dancing with Ice

The sky deck of the Willis Tower in Chicago stands 1,500 feet over the city's streets and is home to a panoramic vista that extends for miles over Lake Michigan to the east and the farmland of Illinois to the west. While the glass floor of the viewing station can induce a dizzying swoon, the history of the area is even more vertiginous. Some 26,000 years ago this deck would have sat 1,500 feet below the surface of an ice sheet that extended along the northern United States through the entirety of Canada. The highest skyscrapers in Manhattan would have been encased in ice every bit as much as those in Chicago. Other glaciers up to 3,000 feet thick extended over Europe and across Asia. During this time, the climate was dry, cold, and harsh. So much water was held by massive continents of ice that sea levels were 420 feet lower than they are today.

Over the ensuing 5,000 years, the ice melted as glaciers retreated to higher latitudes. As they moved, glaciers carved the landscape, leaving valleys in some places and hills and escarpments in others. When the sheet that covered Chicago melted, Lake Michigan and the other Great Lakes formed. Retreating ice carved many familiar points on the globe, from Cape Cod and the Finger Lakes region of New York to Scandinavian coastlines and fjords.

These climatic and geographic changes affected all living things. During ice ages, conifers such as larch, spruce, and fir were common in Illinois, Ohio, and New York. As the ice retreated, deciduous forests composed of oak, ash, and elm replaced them. While the climate was cool, musk ox, lemmings, and caribou, all familiar polar mammals, roamed the plains of the Midwest, along with ice-age species such as dire wolves, woolly mammoths, and saber-toothed tigers. As ice melted, many of these polar mammals retreated north while different species, such as deer, squirrels, and others replaced them. Like human snowbirds wintering in Florida, many species moved south to more hospitable climes during glacial times. Certain kinds of animals found isolated refuges from the ice in the north. Some small patches of land were not covered by glaciers because they sat at altitude or experienced relatively warm microclimates. The creatures that entered these oases typically lived in small groups isolated from other members of their own species. Isolation means inhabitants can evolve traits that are different from their cousins living elsewhere. Genetic isolation is thought to account for the origin of polar bears and for the origin of new species of polar rodents, for example. Isolated refuges are fuel for species to become ever more diverse.

Our predecessors lived and evolved during these convulsive times. Given their lack of agriculture and that their technology consisted of stone tools, even small environmental changes threatened their existence. Like that of other animals, our ancestors' response to extreme climate change was to migrate to more habitable climates, adapt physically over time, and develop new cultural practices and technologies, or die.

Fossil and genetic evidence puts human ancestors in Africa until about 2 million years ago. There were several different species of bipedal ape living in Africa at this time. One of these relatives, the relatively big-brained *Homo erectus*, left Africa and spread across Europe, the Middle East, and Asia. *Homo erectus* developed tools and flourished outside Africa until 108,000 years ago, when the species disappeared. Some anthropologists have proposed that changes to the climate affected its ability to survive.

*Homo erectus* populations that remained in Africa thrived and gave rise to another relative, *Homo heidelbergensis*, by 800,000 years ago. Like *Homo erectus*, this species had a relatively large brain and developed new and ever more complex stone tools and other implements. By 600,000 years ago some populations of *Homo heidelbergensis* migrated out of Africa into northern Europe and Asia, while others remained in Africa. These northern migrants may not have known it, but their world would soon become dominated by ice.

As glaciers expanded and contracted over Europe and Asia during the ice ages, the populations of *Homo heidelbergensis* that had migrated out of Africa developed new technologies. One of them, fire, changed the way they lived, ate, and spent time together. They built temporary dwellings and hunted large animals with wooden spears. And, during glacial periods, they retreated into refuges as smaller groups. Over time, this isolation led to increasing anatomical diversity. Populations of *Homo heidelbergensis* started to look different from one another and a new kind of robust human species appeared by 400,000 years ago: *Homo neanderthalensis*, or Neanderthals. With shorter limbs and a bulky body, Neanderthals were more adapted to cold than other human relatives. *Homo heidelbergensis* populations that remained in Africa followed a different trajectory. These populations were geographically removed from the European ones and evolved separately. By 300,000 years ago, southern African populations of *Homo heidelbergensis* gave rise to modern humans, *Homo sapiens*. By 54,000 years ago, our species left Africa and moved to Europe. *Homo sapiens* coexisted with Neanderthals until 40,000 years ago, when Neanderthals became extinct, leaving only a single kind of human on the planet.

An international team of scientists constructed a massive database containing every occurrence of *Homo heidelbergensis*, *Homo neanderthalensis*, and *Homo sapiens* in the archaeological record—a list of hundreds of sites in Europe, Asia, the Middle East, and Africa. Using one of the world's most powerful supercomputers, they looked to see how the occurrence of each species mapped to the local climate in each location. A surprisingly elegant, yet unexpected, pattern arose. The movement and evolution of the different groups across Africa, Europe, and Asia followed habitat, glacial, and climate changes caused by the orbital cycles that Milanković proposed in 1941.

Ice ages shaped our origins. The populations of *Homo heidelbergensis* and *Homo sapiens* in Africa were affected by the ice ages every bit as much as those living with glacial ice in Europe and in Asia. New ocean currents and wind patterns established by polar ice led to changeable environments in East Africa, the cradle of our species. During this time, our ancestors developed larger brains, new kinds of hunting implements, and different cultural practices. Some anthropologists have proposed that adaptation to these variable environments was a factor in our species' cognitive development. The most recent ice age, which witnessed a peak of ice 26,000 years ago, changed sea levels, giving our ancestors new pathways to disperse across the globe and enabling them to migrate from Asia to North America over the Bering Land Bridge. Humanity—along with art, religion, increasingly intricate tools, and new ways to communicate—arose during times of ice.

Ice sheets continue to loom large in our lives today. Sea levels set by polar ice have defined coastlines, cities, and many of our national borders. Entire island nations exist only because of ice at the poles; the same is true for most coastal cities. Productive fishing areas rely on ocean currents influenced by polar ice. Weather patterns derived from frigid polar regions help create the agricultural belts on which modern society relies. The climate of cities such as Chicago, London, and Beijing are tethered to happenings with Arctic Sea ice and glaciers. Ice has made regions of the globe off-limits for trade routes, extensive human habitation, and extraction of resources such as oil, gas, and minerals. The existence of ice at the North and South Poles is like an invisible hand that has shaped our history and continues to frame our lives.

When David Foster Wallace addressed the graduating class at Kenyon College in 2005, he opened his speech with a story about fish:

> There are these two young fish swimming along and they happen to meet an older fish swimming the other way, who nods at them and says, "Morning, boys. How's the water?" And the two young fish swim on for a bit, and then eventually one of them looks over at the other and goes, "What the hell is water?"

The polar scientist in me wants to replace fish with humans and liquid water with ice in Wallace's parable. We are often blind to our most important realities—in our case, polar ice. Ice is a silent partner in all of our lives. But our activities are changing it—our modes of transportation, food production, and energy use are pumping carbon into the atmosphere faster than it can be removed. In the world that is emerging from these activities, our relationship with the poles will become ever more apparent.

# CHAPTER 7

# ON THIN ICE

A chorus of giggles jolted me from my smartphone. I had been video conferencing with my kids, who were on Cape Cod in Massachusetts, while, unbeknownst to me, three Inuit children had been goofing around behind my back. It was August 2014, and my team had just completed a five-week expedition in northern Ellesmere Island, having had little contact with home. Our return flight to base was interrupted by bad weather, and the plane had to divert to the hamlet of Grise Fiord on the southern part of the island. Home to 160 people 900 miles from the North Pole, Aujuittuq ("the place that never thaws out"), as it is known to residents, is one of the northernmost human settlements in the world. Hundreds of miles from any other town, it sits on the coast of Ellesmere Island with the only access by aircraft or cargo ship in the summer. Sheltering there during foggy days, I whiled away the time pacing its muddy street and exploring the coastline beyond. During these traverses, I had found the best wireless connection in the hamlet at a quintessentially Canadian location—the hockey rink.

My kids and these Inuit children were all between eight and

twelve years old, and had begun to engage each other playfully while I was on the call. Feeling increasingly like a third wheel in my own conversation, I handed the Inuit children my device. They grabbed my phone and scurried to the windows of the community center to show off their town to my kids. The video contact happened at an opportune moment. The annual summer cleanup was happening, and there were polar bear bones, old hunting gear, and other items lying on the ice-and-dirt roads outside the town center. Meanwhile, my children were showing the Inuit children views of Cape Cod, with pine trees bordering a salt marsh and the bay beyond.

The worlds these children were sharing could not have been more different. My children inhabited a green scrub pine forest with the blue hues of a bay in the distance while the Inuit children dwelled in an expansive white-and-taupe world of snow, ice, and rock. As the kids were discovering differences between their landscapes, it was clear something deep connected them. Their lives are intertwined in ways they could not see.

## The Idea of a Lifetime

Willi Dansgaard was born in 1922 and trained in physics at the University of Copenhagen. Like many Danes, he longed to go to Greenland after earning his degree. In 1947, he leapt at an opportunity to work on Disko Island in West Greenland, measuring the geophysical properties of the Arctic. Dansgaard and his wife, Inge, moved to a small village in Greenland, only to find there were not enough supplies with which to build a house. After they spent nine months living with another couple, a shelter was eventually built, only to burn to ashes two weeks later. Losing many of their personal effects, the couple survived off of a local cache of goods that fortunately included an excess of rye, oats, and chocolate. Having eaten chocolate at every meal during this time, Dansgaard reminisced later that he "could not bear to see the stuff for years afterwards."

With a stable hut eventually secured, Willi and Inge read to each other during the long polar nights while Dansgaard worked on his scientific devices and measurements during the day. Like many visitors to the region before him, he was transformed by his polar experience. He was to later say that during this time with Inge he was "smitten with Greenland for life" and permanently changed by "its forces, its bounty, its cruelty and, above all, its beauty."

Dansgaard returned to Copenhagen in 1949 with a yearning to go back to the Arctic, but he needed to make a living first. Postwar Copenhagen of the late 1940s did not offer many opportunities. Dansgaard secured a government post as a librarian in a national collection but soon soured on the drudgery of the job. He didn't have to wait long for a new gig. The ground was shifting for research—the Marshall Plan funds that the United States provided Denmark to rebuild after the war helped the nation purchase new scientific infrastructure. The geological institute in Copenhagen used the influx of cash to purchase a mass spectrometer, a new device that could measure the atomic composition of different substances. Seeing an opportunity to work in science instead of being a bureaucrat, Dansgaard gave up the safety of his government job to become a technician on the machine. He supported himself with the fees people paid to use the device. The new post was, in fact, an excuse to play with the mass spectrometer during off hours to be able to pursue his own scientific ideas.

Freed from filing books, Dansgaard used the mass spectrometer to explore the physics of different materials, most importantly water. He hadn't lost his passion for Greenland, and he thought that if he could develop a way to assess the molecular composition of ice and liquid water, he might be able to understand how the glaciers of the Arctic formed and evolved over time.

In thinking about ice, mass spectrometers, and polar regions, Dansgaard hatched an idea. He knew that the element oxygen existed in two major atomic forms—a heavy one, known as $^{18}O$, and a lighter one called $^{16}O$. He guessed that the two forms of oxygen behaved differently in water depending on the temperature. He theorized that when water evaporates from a liquid to a gas, the gas would be enriched in the lighter form of oxygen. The obvious assumption is that lighter substances are more likely to float into the air. Likewise, he conjectured that when liquid water condensed from vapor, the liquid holds more of the heaver version of the element. Dansgaard envisioned that this process could give an estimate of temperatures. Higher temperatures provide more energy to move the lighter form of oxygen from water to vapor, and lower temperatures mean more of the heavy form of oxygen stays in liquid. This was bold thinking—if Dansgaard was correct, then the ratios of the heavy and light forms of oxygen in water might serve as a thermometer to indicate ancient temperatures in a sample of liquid water or ice.

Having a mass spectrometer at his disposal, all Dansgaard had to do was to find different waters to test the soundness of his idea.

Lacking extravagant resources, Dansgaard's first experiment was close to home, and low-budget. His plan was to look at the rainwater that fell in his garden at different times of the year. If his theory was correct, the oxygen atom composition of rainwater should show a relationship to the temperature at which it fell. His first collection device was as simple as his idea: an empty beer bottle with a funnel at its mouth, placed in the middle of his lawn. He set up a number of these small contraptions in his backyard in Copenhagen each time the forecast called for rain. Dansgaard then got lucky—the year of the experiment was particularly wet, and he was

Willi Dansgaard with an ice core

able to collect rainwater during storms throughout the year. After months of collecting water during rainfalls when the climate was warm and then cold, there came the moment of truth—measuring the oxygen atoms in his mass spectrometer.

The ratio of the two forms of oxygen conformed nicely to the temperatures at which the rain fell at different times of the year. Dansgaard's analysis of the waters inside his backyard beer bottles would change our understanding of polar regions and the effects of climate change across the Earth.

Dansgaard had a new mission. His technique meant that he could measure ancient temperatures in water by looking at ice. This realization meant he could now explore the climate history of the polar north. What stories about ancient temperatures of the planet are held inside the ice of the Arctic? How did climate change over time?

Once word got out about his technique, Dansgaard received ice samples from scientists who were collecting all over the world—mountain glaciers, oceans, streams, and polar ice. But the puzzle that motivated Dansgaard—a continuous history of ice on Earth—remained elusive. All he was receiving were pieces of information; a chunk of ice here, a collection of water there. A cohesive picture had yet to emerge from any single place. Dansgaard needed a *continuous* section of ice that showed temperature changes over a long period of time.

Dansgaard's big break came from an unlikely place. In the 1950s, Cold War politics led the Pentagon to consider a military base, with nuclear missile silos, in the area closest to Russia: the far northern polar regions. The plan was to build an entire military base under the ice in northern Greenland. Known as Camp Century, it had quarters for enlisted men and officers, dining facilities, and a chapel. The base extended hundreds of feet deep in the ice. To test the ice's suitability for the base, the U.S. military had devised new technology to drill a core 4,560 feet deep from the surface of the ice to its contact with the bedrock below. This core was exactly the kind of continuous section of ice that Dansgaard was seeking.

The generals of the U.S. military were initially unwilling to let Dansgaard have access to its ice core for now-obscure national security reasons. But as has happened many times in polar research, geopolitics became ever more important. The United States wanted a stable presence on Greenland, and good relations with the Danes were essential for this relationship to grow. While the military refused to allow Dansgaard direct access to Camp Century, they provided access to the cores stored in a remote lab. In 1964, Dansgaard was able to analyze a section of glacial ice that extended roughly 4,560 feet, representing 100,000 years of time. Dansgaard couldn't have hoped for more—the core was from the time when glaciers covered Europe, Asia, and North America to when those glaciers melted and receded, starting 15,000 years ago.

Dansgaard's initial result was satisfying, if not surprising. The end of the last glacial era, around 21,000 years ago, saw a period of

"Main Street" in Camp Century under construction in the ice

warming until the glaciers receded to their current position 11,000 years ago.

But the closer he looked, the more Dansgaard found some big puzzles in the ice.

Dansgaard, like the entire scientific community, thought that the times when glaciers were at their maximum size were boring affairs. These so-called glacial periods were considered to be stable and cold intervals on Earth. But Dansgaard's oxygen atoms were showing something else in play. He knew ice cores have dust and other particles inside them that form very fine bands. These layers develop due to changes in wind and temperature over single years or even, in some cases, individual seasons. For example, the different wind patterns in fall and winter are reflected in the amounts of dust in the ice core. Captured as layers, these intervals gave

Dansgaard a very detailed picture of changes to the ice over time. He could now map those changes precisely.

Dansgaard's first attempt at chronicling temperature changes led to more questions than answers. He found short intervals, lasting a few decades, when the polar region warmed by as much as 10 degrees Fahrenheit. These balmy times were followed by a slow cooling that lasted a century or more. The temperature changes were orders of magnitude faster than anybody had considered possible. But the more Dansgaard refined his technique, the more confidence he had in his result. Earth's polar climate can change very quickly. When Dansgaard published his Greenland core results in 1964, the impact was minimal. The scientific community largely ignored his findings, assuming they might be an artifact of Dansgaard's approach or some imperfection in the core itself.

In 1977, Denmark and Switzerland developed a collaboration on Greenland, analyzing drill cores almost 2,400 feet deep. Working with Swiss scientist Hans Oeschger, Dansgaard analyzed the chemistry of the ice in this new, more refined core. With a deeper and more precise core, Dansgaard's earlier discovery became scientific fact—rapid swings in temperature became impossible to ignore. In the years since, wherever scientists have analyzed the history of Arctic ice over the past 100,000 years, they've uncovered intervals of rapid warming even when glaciers were at their greatest size. Rapid changes to the polar ice were a fact.

Cores from elsewhere on the planet also began to show more ancient upheavals in the Arctic. In the mid-1980s, German oceanographer Hartmut Heinrich was studying cores in the ocean floor that covered the same 100,000-year interval as Dansgaard's Greenlandic ice. Heinrich sampled layers of mud at the bottom of the Atlantic Ocean in an effort to understand changes to the ocean since the last glacial period. Heinrich had already known of the wild climate swings described by Dansgaard and Oeschger, but even that did not prepare him for what he saw.

Typically, cores of the ocean floor consist of regular layers of muddy sediments that contain small, single-celled marine organisms, such as foraminifera and ancient plankton. This gradual and continual action means that ocean sediments become a detailed record of changes over a long period of time, much like the ice cores that interested Dansgaard. Most of Heinrich's core held true to his expectations. But eleven layers were utterly different—they contained rock debris. Each held broken rubble not naturally present in any ocean on Earth. Moreover, the sediments looked like crust fragments found in eastern Canada. Guessing this layer of debris was transported to the ocean floor from a site in Canada, Heinrich proposed the only mechanism he knew that could account for this movement: ice. He conjectured that thousands of years ago, icebergs broke off of glaciers in Canada, rafted into the ocean, and deposited rocky debris that settled to the ocean floor as they melted. Heinrich envisioned that this armada of icebergs was set in motion by the breakup of glaciers.

The more scientists analyzed ice cores and the seafloor, the stranger these iceberg-glacier breakup events became. The glacial breakups, involving the separation of hundreds of large icebergs from an ice sheet in northern Canada, happened over the course of *less than 250 years*. And just as rapidly, each one of these events elevated global sea levels between 3 and 6 feet.

By coring through polar ice and ocean muds, Dansgaard, Oeschger, and Heinrich discovered the Arctic has tipped swiftly from warmer to colder climates while ice sheets broke into the ocean and changed sea levels. The capacity for very rapid changes is a natural part of the system of ice, climate, and ocean currents.

The planetary history inside the cores showed dramatic changes without people exerting any effect on global climate. But what happens when you add human industry to the mix of ice, rubble, and oceans?

## Visible to the Naked Eye

In contrast to the Arctic, Antarctica was seen as a stable bulwark against changes in ice. The vast continental ice of Antarctica sits at an average altitude of 9,800 feet. Only dramatic warming of the air could feasibly change Antarctica in any appreciable way. This chain of logic leads to the idea that Antarctica is a firewall from the kinds of sea-level changes that can happen from a melting Arctic. Some theories even envisioned Antarctic glaciers only *growing* with a rise in temperatures. As recently as 2001, the United Nation's Nobel Prize–winning Intergovernmental Panel on Climate Change (IPCC) declared, "The Antarctic ice sheet is likely to gain mass because of increased precipitation." The theory was that a warming climate causes more water to evaporate from the ocean and settle as snow that falls on Antarctica. Ultimately, both factors would lead to the growth of the ice sheet over time. IPCC reports come out every five years, and they drive industrial and social policies of nations around the globe. Antarctica was not a place to worry about in a warming world. Or so the thinking went.

But rocks were revealing something vastly different.

After a stint as a radio operator in World War II, John H. Mercer trained as a glaciologist and held various research positions across the globe. His life changed dramatically in 1960 when he joined Ohio State University's renowned Byrd Polar and Climate Research Center. Founded by an endowment provided by Admiral Richard Byrd, it became a major center of research on polar glaciers, with

John Mercer (clothed) in Antarctica in 1965

scientific luminaries teaching or training there. Mercer was a keen observer of rocks and an avid field geologist who had famously eccentric habits in the field; he reputedly did some of his fieldwork in the nude on (relatively) sunny Antarctic days. Knowing the conditions in these areas, I presume he was wearing boots, at least.

By the late 1960s, Mercer was among a number of geologists tasked with understanding the history of Antarctica. He set off to map the sequence of rock formations alongside a glacier that lies about 1 mile above sea level and 600 miles from the South Pole. This glacier, known as Reedy Glacier, lies in West Antarctica, which is separated from the larger East Antarctica by the Transantarctic Mountains. Mercer noticed something strange in the geological sequences along Reedy Glacier's rocky walls. There, layers of sediments were exposed as finely grained shales that were typical of those found at the bottom of lakes. Inside these shale beds lay rocks and boulders of different sizes that seemed to be randomly scattered about. In any other setting, this hodgepodge of stones of varying shapes, sizes, and composition would be interpreted as debris from glaciers that, having scooped up the landscape for miles as they moved, deposited the remains as they melted. The implication seemed crazy. At some point in the past, the place held lakes that were formed by melting glaciers. Atomic analysis suggested that the lake sediments and glacial debris were 120,000 years old.

The Reedy Glacier sits about 1 mile higher than the coastal glaciers of West Antarctica. Knowing this fact, Mercer made a conceptual leap: if this high-altitude site held lakes instead of glaciers 120,000 years ago, then clearly the regions of the West Antarctic Ice Sheet, closer to sea level, had liquid water instead of ice as well. Mercer's logic implied that the entire West Antarctic Ice Sheet was nonexistent at this time. He proposed that during warm periods, the West Antarctic Ice Sheet melts completely, only to re-form during cold ones.

The implication was as profound as it was unsettling. The ice of West Antarctica is nearly one and a half miles deep in places. It has a surface area about the size of Alaska and holds 7.2 million cubic *miles* of ice. If that ice melted and entered the ocean, sea levels would rise about 14 feet worldwide.

Knowing the implications of his idea, Mercer then looked inside geological layers around the globe to see whether sea levels had been any different 120,000 years ago. Some of the sites with the best sediments to address this question lay in the Pacific. Stands of rocks containing fossils of marine organisms found along the coasts in Indonesia suggested that the sea levels at this time were almost 14 feet higher than they are today. To Mercer, the connection to Antarctica was inescapable: the amount of sea-level rise as measured in Indonesia matched the volume of water that would be added if the West Antarctic Ice Sheet melted completely and entered the sea.

If the Antarctic was supposedly so stable, how could such a vast ice sheet melt so completely?

During the International Polar Year of 1957, the one that launched the discussions for the Antarctic Treaty, nations of the world got together to perform polar research. Teams from Russia, the United States, Britain, and other countries set off to map and study different parts of the poles. One team traversed the West Antarctic Ice Sheet and took cores of the ice. At the time, little was known about the bedrock underneath the glaciers, let alone the ice of the glaciers themselves. As the teams hit bottom on their deepest cores they were able to decipher what the land looked like below. The West Antarctic Ice Sheet sits in a bowl of bedrock, most of which lies below sea level. The ice sheet is so heavy that it depresses the continent to the extent that portions of West Antarctica lie almost 2 miles below sea level. By comparison, Death Valley, one of the deepest spots on Earth, lies 280 feet below sea level. The depth of bedrock under the ice gave Mercer an idea of why the West Antarctic Ice Sheet is so susceptible to rapid melting.

If the bedrock under the glacial ice sits below sea level, then ocean waters can seep underneath. Because this seawater holds heat, it will undermine the glaciers, thinning them from below. The seawater will then expand ever inland as the ice above melts. But what happens at the Antarctic coast does not stay at the coast: ice streams that flow from the center of the continent are held back from entering the ocean by coastal glaciers. Like a dam holding water, ice at the coast holds continental ice back from flowing into the sea. These coastal glaciers form cliffs along the sea. Projecting from them are shelves of ice that float over the ocean for miles. Mercer and his colleagues envisioned that warming oceans would melt the coastal glaciers from below and cause the floating ice shelves to collapse. Then, the ice dam would collapse and ice would flow and spill ever faster into the ocean. To Mercer, this is what led to the loss of the entire West Antarctica Ice Sheet 120,000 years ago. He was sure this could happen again.

Mercer described his ideas in conference proceedings in 1968. But moon landings and the Vietnam War held more attention than obscure theories about Antarctic glaciers. Instead of setting off alarms, Mercer's paper sparked little interest. There wasn't even a question or discussion about it at the conference.

If Mercer didn't care about wearing clothes on Antarctic glaciers, he certainly didn't mind negative opinions or indifference from colleagues. Doubling down on his idea, he published a new and fuller version of his paper in 1978. He had realized that the production of greenhouse gases in the atmosphere can raise temperatures to the same levels that had caused the melt of the entire West Antarctic Ice Sheet 120,000 years ago. By Mercer's calculations, the world was set for a 14-foot sea-level rise that could happen catastrophically in the next forty years as the ice shelves holding the West Antarctic Ice Sheet break apart. This change could displace more than 1 billion people and inundate or alter every coastal environment on the planet. Mercer ended his paper with the plea: "One of the warning signs that a dangerous warming trend is under way in Antarctica will be the breakup of ice shelves on both coasts of the Antarctic Peninsula, starting with the northernmost and extending gradually southward. These ice shelves should be regularly monitored by LANDSAT imagery."

Mercer was denounced as an alarmist. An editorial in a scientific journal said his idea read like a B movie. He struggled to get grants to support his research. Much of the scientific community still envisioned Antarctica as a world of stability, not a place that could change quickly. Dansgaard, Oeschger, and Heinrich's rapid changes were in the Arctic. A stable sheet of ice over the Antarctic continent could never behave that way.

While Mercer's ideas failed to gain traction in the larger scientific community, a small group of other scientists picked up the thread. One of them, Terry Hughes, an Ohio State colleague of Mercer's, noticed that two of the main glaciers that hold back much of the volume of the ice sheet were in a precarious state. The ice shelves appeared to be small relative to the glaciers themselves. In addition, the adjacent ocean waters didn't hold much ice, and they looked to be warming. To Hughes, these features meant the West Antarctic Ice Sheet was at risk because of the weakness and precarious position of these two glaciers, Pine Island and Thwaites, and the surrounding warm water. He published this idea in a scientific paper with the title "The Weak Underbelly of the West Antarctic Ice Sheet" in 1981. Hughes's work was met with the same indifference his colleague Mercer's had experienced.

But events have a way of changing even the most deeply held scientific opinions.

## Witness to a Breakup

In 1994, Pedro Skvarca, a professor at the Argentine Antarctic Institute, published a relatively obscure paper on satellite images of the so-called peninsula region of Antarctica. This region, as the name implies, extends as a 750-mile-long projection north from the central body of the continent, toward South America. It is an extension of the Transantarctic Mountains and borders the Weddell Sea. Many people know it because it hosts the vast majority of tourist cruises that probe into the continent. If you see a picture on social media of somebody in a Zodiac, wearing a parka, with penguins in the distance, it most likely was taken in the peninsular region of Antarctica or along the coast of the Weddell Sea. The peninsula has scientific renown because along its length lie shelves of ice that float on the ocean waters. They are connected to glaciers inland and they effectively serve as barriers that prevent the ice from the interior of the peninsula from entering the sea.

Skvarca wanted to see what changes were happening to one portion of this region of ice, known as the Larsen Ice Shelf, which lies plastered against the land of the peninsula on its western side. The shelf is huge and is divided into five different regions that run north to south and are called Larsen A through Larsen E. Skvarca wanted a picture of how the shelf is changing over time, but he had only crude data at his disposal because there were relatively few satellites taking pictures of the Earth below. Skvarca was able to wrangle images from the KATE-200 camera on a Soviet satellite taken in 1975 and others from the American Landsat from 1986 to 1989. The images were not strictly comparable because they were taken under different conditions and with different equipment, but the conclusion was inescapable: in the decade covered by the two sets of images, the ice shelf had developed rifts and clefts and receded across its extent. Skvarca saw the Larsen Ice Sheet shrinking and likely breaking up, but he ended his paper with a plea for more refined data. He needed more precision and measurements in real time.

Skvarca didn't have to wait long. Argentina runs Matienzo Station, a tiny research base sitting on a volcanic island 30 miles from the coast of the peninsula. Surrounded by Larsen A, a shelf of ice the size of Los Angeles, the base typically houses roughly ten researchers. In February of 1995, the crew started to hear strange noises and see new pattens developing in the ice every morning. Gurgles emanating from within the ice and ponds forming on its surface suggested that it was melting. The situation got worse every day. At night the crew heard earsplitting crashes and booms that felt like thunderstorms. New rifts and cracks emerged by the end of each working day. Then, one day during lunch, their world exploded. A roar like a tornado led them outside their hut. They witnessed the ice shelf shattering into pieces. What was once a solid shelf, stable for hundreds of years, broke apart within hours. The crew, who had just witnessed more changes to the polar landscape than any others in recorded history, were soon evacuated. A satellite assessment of the breakup revealed that Larsen A had simply burst apart and was carried away by the sea in a matter of days.

The speculations about Antarctica being a stable buffer of change melted away as fast as the Larsen A shelf. Eyes were now on the other Larsen ice shelves—B, C, D, and E. Would they start to disappear too?

NASA's Terra satellite is the size of a small school bus and carries five different cameras to record changes on the surface of the Earth. One of them, known as MODIS, has such a wide field of view that it can scan every point on the Earth every two days. On January 31, 2002, images showed a series of parallel blue streaks, miles long, forming rapidly on Larsen B. The blue streaks told of water on the surface of the ice. The fact that they were in parallel lines suggested that the surface was melting and the water was pooling and flowing inside cracks and crevasses on the ice shelf. Images were not available every day because the cameras weren't continually trained on the site or clouds obscured the view. Waiting a few days, researchers obtained another MODIS photo that showed the blue streaks had disappeared. Rather than celebrate, they realized that the cause of the disappearance was likely due to the water in the crevasses leaking out from holes in the bottom of the ice shelf. Something urgent was happening to Larsen B, echoing the melting of Larsen A in 1995. The next clear satellite image wasn't available until February 17, but it showed that the leading edge of Larsen B had retreated by 5 miles. By March 7, much of Larsen B had disintegrated, leaving a blue-colored slurry where an intact ice shelf had been a few months before.

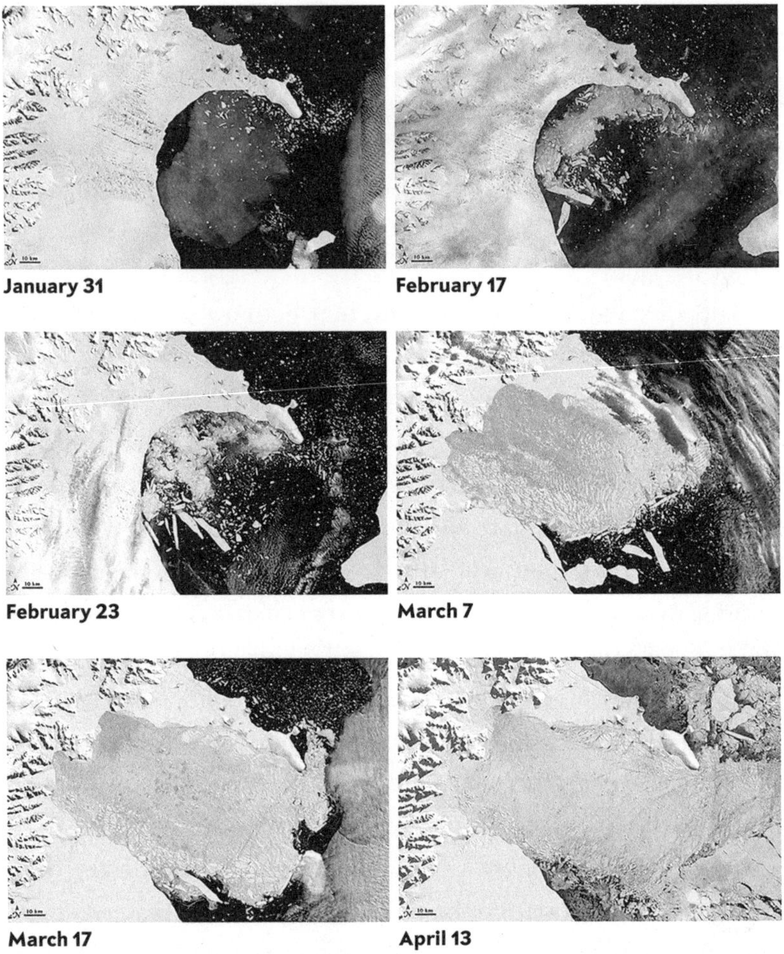


The breakup of the Larsen B Ice Sheet in spring 2002

The area of Larsen B that melted in 2002 was two times that of Larsen A in 1995. An ice shelf the size of Rhode Island was now gone and melted into the sea in a matter of weeks.

The disappearance of these vast ice sheets coincided with periods of warm summers. Warming air and oceans melted the shelves from the top and the bottom, causing them to shatter and cleave into numerous icebergs.

While the disappearance of these ice shelves is alarming and reveals that ice can retreat and disappear far more rapidly than scientists had previously thought, nobody was worried about immediate changes to global sea levels or life at coastlines farther north. Ice shelves mostly float over the ocean, like an ice cube in a glass. A melting ice cube doesn't change the volume of the water in a glass any more than a melting ice sheet would an ocean.

But these facts hide a deeper problem. If ice shelves behave, as Mercer and his colleagues envisioned, like dams that hold back glaciers on land from entering the sea, then their disappearance could cause much more ice to enter the ocean. If one the size of a large city can break up in a matter of days, then the floodgates could open for an enormous amount of continental ice to slide into the sea.

Since the melting of Larsen A and B, as well as other ice shelves on the peninsula, teams have looked at the movement of the ice streams that carry ice from the interior to the coast. As predicted by Mercer and other glaciologists, the loss of the shelves led to an increased flow of ice into the sea. The rate of flow increased from fourfold to eightfold. That amount of flow is not a huge problem for sea-level rise because the peninsula is a thin strip of land with no major ice sheets inside.

That's the good news: the Larsen ice shelves do not hold back much ice. The bad news is that there are other ice shelves that hold back enough ice to raise global sea levels by over 16 feet in a single century.

## Fingers in the Dyke?

The collapse of ice sheets the size of large metropolitan areas in a matter of days meant that Mercer's and Hughes's ideas moved from speculation to unsettling possibility.

A series of six glaciers, including Pine Island and Thwaites Glaciers, hold back ice streams of the West Antarctic from flowing into the ocean. Thwaites is the largest of these glaciers that serve as barriers. If Thwaites melted entirely, then global sea levels would rise about 2 feet. But if the shrinkage of Thwaites led to the collapse of the entire West Antarctic Ice Shelf, then the ice streams would flow into the ocean and sea levels might rise as high as 16 feet. But are those extreme scenarios possible, or even likely, with current levels of global warming?

Understanding these glaciers involves years of planning and coordinating the expertise of diverse teams of carpenters, drillers, pilots, mountaineers, and scientists. Thwaites and Pine Island Glaciers are almost 1,000 miles from the nearest base. A small, temporary settlement is built on the remote ice where everything, from snowmobiles to helicopters, is brought in for a three-month summer season, then removed for the rest of the year. But even planning these encampments can often take years of reconnaissance to test whether aircraft can land on the ice and whether a camp can be established safely. Life there is at the end of a long logistic chain—a problem with aircraft or some of the mechanical gear can shut the entire program down. Key personnel may fail the medical tests necessary to get approved to work on the continent. A single problem, out of this sea of potential ones, can derail finely planned fieldwork for an entire season. Resilience, coupled with ingenious technology, is the watchword for successful science on the ice.

The challenge to studying the largest ice sheets on the planet is one of scale. The Antarctic Ice Sheet has more surface area than the entire continent of North America, whereas the one in Greenland is about the size of the state of Texas. Understanding the behavior of glaciers means observing from space, on the ice, and even below it.

All of these efforts seek answers to the following questions: How are these glaciers changing? How fast is it happening? What do these changes say about the future of sea levels?

With renewed interest in West Antarctica in the 1990s, Eric Rignot and his colleagues deployed radar interferometry to measure the motion of the ice at Pine Island and Thwaites. Recall that this approach allows him to see movement of the ice at the incredible resolution of 1 millimeter every time the satellite's orbit passes overhead. The Landsat images that Skvarca used, meanwhile, were merely static views of the surface of the ice. Rignot and his colleagues even devised clever techniques to see where the glacier made contact with the bottom of the ocean. Glaciers undergo an important change as they enter the sea. While moving over the continent they cover the bedrock below. When they enter the ocean, they form a floating shelf of ice above the water. The line where the ice changes from sitting on bedrock to floating on water is of special significance. Changes in the location of this line reveal whether a glacier is shrinking or growing. When the glacier retreats, this feature moves inland, while a growing glacier will have this line moving into the sea.

Rignot and his colleagues were able to estimate the position of these lines for different glaciers in West Antarctica. The part of the glacier that sits over water bobs up and down with each tide. The part of the glacier that sits over land does not. The place where the bobbing stops is the point where the glacier ends its contact with the land and extends as a shelf over the ice.

From the moment he first saw the data, Rignot was uneasy. Invited to present his results at an Antarctic glacier conference in 1997, he was so nervous that his viewgraphs were upside down during the presentation. After his talk, his fears were realized: "A lot of the crowd was not very warm or sympathetic to the news," he told me.

His images showed that the contact of Pine Island Glacier with the edge of the continent was retreating 1.2 kilometers per year between 1992 and 1996. Not only was the glacier retreating 20 meters per day, it was also changing on the inside. The image data revealed the glacier was thinning more than 3 meters per year as it retreated. Gigatons of fresh water were entering the oceans as the glacier shrank. The glacial community had a new quest: understand what is happening to Pine Island and Thwaites Glaciers. Mercer and Hughes weren't alarmists; they were sages.

Rignot's data came from outer space. More, recently, new views of the glaciers have come from inner space. In 2019, an expedition set off on the National Science Foundation's research vessel *Nathaniel B. Palmer*, with a remote submersible designed to map the seafloor. Scientists launched the craft and piloted it from the surface as it mapped 8 square miles of seafloor with specialized cameras and sonar. Their objective was to examine the seafloor directly in front of the Thwaites Glacier. These data allowed them to see how the glacier had moved in the recent past by observing the scrapes it made on the ocean floor.

When the submersible returned to the ship, one of the leads of the research program said, "None of us could explain what we were seeing." Nobody had looked at the ocean floor underneath a glacier before. Series of parallel ridges about 10 to 70 centimeters tall extended along the seafloor, regularly spaced from one another almost as if somebody had carved them. The ridges extended from the end of the glacier into the seafloor and ran parallel to the edge of the glacier. This is a part of the ocean that lacked any currents that could feasibly have sculpted these features. The only explanation was that these ridges were made as the glacier bobbed and receded as it retreated. This observation was confirmation of the glacier retreat Rignot was seeing from space. The spacing of the ribs reflected the timing of the tides: when the tide was high, the ice rose; and when the tide was low, it descended and pressed the ocean sediments to make a rib. Then the cycle repeated the next day. The ribs on the ocean floor formed a kind of ocean tidal chart: the height and distance between ribs matched the daily and monthly cycles for the area.

The marks on the ocean floor, and the way they matched the tides so perfectly, gave the researchers a kind of clock to measure how fast the glacier is retreating. By counting ribs and measuring distance, they found that the glacier had retreated 2.2 kilometers over the course of five months at some point in recent history. That rate is twice as fast as the retreat measured by satellites today. These marks on the seafloor were likely made during the past two centuries. Thwaites Glacier, it seems, is capable of very rapid retreat and is extremely sensitive to changes in the environment around it.

Eric Rignot and Sridhar Anandakrishnan are part of an international team trying to learn what is happening to Thwaites Glacier as it meets the ocean waters below. From 2014 to 2020, scientists from Cornell University and Georgia Tech built an 11-foot-long, 10-inch-wide torpedo-shaped robot to slither through drill cores and maneuver underwater beneath Thwaites Glacier. Nicknamed "Icefin," the robot carries cameras, laser ranging systems, and instruments to measure properties of the water and to send data back to the surface.

In 2020, the team drilled a 2,000-foot core and slid the probe below Thwaites to provide the first picture of the ice. The images were both surprising and complex. The glacier's underside is relatively flat and has been thinning smoothly and steadily. But the melting rate of this part of the glacier is slower than computer models predicted. Freshwater from the melting glacier interacts with the warm ocean water below to form an unexpected insulation barrier. While that seems to be good news, rapid melt is happening in cracks and crevasses where warm waters flow inside the glacier. Most of the shrinkage and structural change to the glacier is happening at its weakest points. While the dynamics between ice

Icefin under the ice

and liquid water are as surprising now as they were to Rendu, Forbes, Tyndall, and Kelvin, technology has the power to answer the important questions. A new generation of Icefin is in the works to make more refined measurements in the coming years.

Not only are the glaciers that support the West Antarctic Ice Sheet receding and getting thinner, the ice shelves that buttress them are melting. Cracks, fissures, and ice pools, not much different from those scientists saw in the Larsen B collapse, are appearing over the shelves. By every measure the ice shelves that support Thwaites and Pine Island are thinning much like the Larsen ice shelves that collapsed so suddenly in the 1990s.

Antarctica, as a continent-sized field of ice, stays relatively colder than the rest of the world. In fact, it is witnessing the slowest temperature rise on Earth. At first, this fact might also seem like good news for places like West Antarctica. But the devil lies in the details—or, more precisely, in the relationship between the air and ocean waters. As we saw earlier, Antarctic ice typically doesn't melt from the top—it melts from below, where the ice meets the ocean. As the oceans warm, Antarctica's surface temperatures stay relatively cool. This disparity in temperature amplifies a wind current, the so-called westerlies that circle the southern part of the globe. By pushing the water, these winds accelerate the famed circumpolar current that swirls around Antarctica. This current carries 170 times more water than all of the Earth's rivers combined. And with that water comes the heat it holds. The change in the circumpolar current brings ever-increasing amounts of warm seawater to the coast of Antarctica. These warm waters cause the coastal glaciers to melt from below. As they get undercut, the glaciers fragment and collapse into the sea like a sand castle in an ocean tide.

The chain of logic for how rising temperatures affect the glaciers of West Antarctica is: rising temperatures change trade winds, which alter ocean currents, which drive warm waters underneath the glaciers at the coast. These warm waters undermine the ice shelves and coastal glaciers that have prevented the ice streams from the interior from entering the sea.

Shrinking glaciers mean rising seas. That part is obvious. But how much, and how fast, waters may rise depends on other factors.

## The Land Also Rises

Born in 1821, Charles Francis Hall spent most of his life starting small businesses, including two newspapers in Cincinnati. Through his writing he developed an obsession with all things polar and became determined to meet with Inuit and to learn the fate of the Franklin expedition. A giant bear of a man with a bushy beard, Hall had a natural charisma that could both inspire friends and, ultimately, provoke enemies. At age thirty-nine, after living an urban lifestyle, Hall searched out the American benefactor of polar expeditions Henry Grinnell and convinced him to make introductions to whalers and other ship captains for passage north.

After gaining support to make a first Arctic trip, Hall went on to live with Inuit, learn from them, and apply their traditional knowledge to problems of geography and history. He seemed more of a lone wolf than an expedition member or leader. He shot and killed a colleague on one of his early expeditions over a disagreement about working with Inuit. From that tragedy he went on to lead one of the greatest expeditionary debacles of all time. Funded by the U.S. Congress in 1871 to attain the North Pole, his trip devolved into misadventure and ultimately near warfare among different factions of his crew. Hall was rumored to have died of apoplexy during the trip, but his biographer, after traveling to Greenland to exhume his body years after his death, revealed that Hall had been consuming arsenic for the last two weeks of his life. Given his poor leadership and fragmented crew, the evidence that they had been slowly poisoning him may not be entirely surprising. It also didn't help that the doctor who was tending to him in his final weeks was a passionate rival for a love interest back in New York.

While Hall had little talent for expedition management, he was far ahead of many American and European peers in realizing the extent to which Inuit understood the Arctic world. Rather than run expeditions in parallel to them, he sought to learn from their millennia of experience with the landscape.

One of the things Hall learned was that the beaches in the Arctic seemed to have been rising over history. The beaches were expanding as the Arctic coastline rose above the water. As he said, "Innuit all think this earth is covered with water. Did you ever see such stones as clams and things that live in the sea up in the mountains?" As Hall came to realize, Inuit were seeing their land elevate, so former beaches were rising into hills. They were finding shells of clams on hills, far from the shores that made up their usual habitat.

Glacial ice the size of entire countries has an enormous mass, enough to cause the land below to sink. Continents float on the deep mantle inside the Earth, so when ice covers them, they sink. On the other hand, when the ice melts and disappears, the land rebounds and rises. Since the mantle is very viscous in places, the rebound can take thousands of years.

The last ice age ended 12,000 years ago when the ice sheets that covered much of Europe and northern North America receded. During those times, the continents sank under the weight of the ice. With the ice gone, the land underneath rose, and because that process is generally slow, some places are still rebounding and rising.

Glacial rebound is a natural part of the maritime history of Nordic countries such as Sweden and Finland. There, harbors retreat from the water over decades. In fact, this process creates a bit of a legal quagmire. Places that were formerly covered by ice are now rebounding so much that new land is being created. Finland even has a new land law to determine who owns the land that appears as coastlines rise.

Glacial rebound is thought to progress extremely slowly, taking hundreds or thousands of years to have a major impact. But rates of rebound are very local. In some places, the mantle is a viscous soup. Here, when the ice retreats, the crust will elevate slowly because the syrupy mantle expands slowly. In other places, the mantle is more like a watery fluid. In these places, the crust can rebound quickly after ice recedes. By some estimates, this rebound can happen over a human lifetime—less than a century.

Seismic analyses of the mantle beneath West Antarctica reveal the complexity of what happens when ice melts. There is no doubt that the ice is melting fast and melting in a way that, unchecked, could lead to an armada of ice from West Antarctica entering the ocean. But the mantle beneath Thwaites, Pine Island, and other glaciers in this part of Antarctica is not viscous. This means that the crust, including the seafloor, can rebound relatively quickly as the ice recedes from the surface. In the most optimistic scenario, a rise in the land as the ice recedes can slow the movement of glaciers into the sea, almost like elevating the lower end of a sliding board to slow down, or stop, a descent to the ground.

This effect offers hope that West Antarctica is more stable than a simple analysis of its ice alone would suggest. But a recent analysis that looked at the combined rates of ice loss and land elevation offered this assessment in academic prose: "Results show that on multicentennial-to-millennial timescales, spatial variability in solid-Earth deformation plays a significant role in promoting the stability of the West Antarctic Ice Sheet (WAIS). However, WAIS collapse cannot be prevented under high-emissions climate scenarios." But these kinds of predictions, while informative, are only as good as the assumptions underlying them.

Given the complex interactions between ice, oceans, and land, predicting the behavior of the system is both urgent and challenging. One way to untangle this knot of relationships comes from peering deep into the ice and into the past.

## Out to Sea

John Mercer revealed how the collapse of ice in West Antarctica 120,000 years ago led to a dramatic rise in sea levels. But could that interval have been a one-off? What are the lessons gained by looking further back in time at the long-term relationship between temperatures, ice, and seas?

As we've seen with the work of Dansgaard, Oeschger, and Heinrich, cores in the ice and seafloor can be extremely precise, revealing histories that can even record seasonal changes hundreds of thousands years ago. The most detailed records from cores extend to about 400,000 years ago. When changes in temperature, sea levels, and the amount of carbon dioxide in the atmosphere are plotted onto the same graph, the rise and fall of each variable are intimately linked; changes in one are linked to changes in each of the others. Temperatures have wiggled up and down about 6 degrees Celsius during this time. With the changes in temperatures came the rise and fall of sea levels. The fact that sea levels rise and falls along with temperature surprised nobody. But the data from the past hold something more revealing.

As temperatures ranged over 6 degrees Celsius for the past 400,000 years, sea levels varied by almost 360 feet. When scientists plot temperature against sea-level rise over the last 400,000 years, they get a straight line—for every 1 degree Celsius of temperature rise, there was nearly 60 feet of sea-level rise. But if we look at the current situation on that graph, there is a disconnect: For our current global temperature, we sit at a relatively low level of ocean, based on data from the past. The position of this point on the graph means that as the globe adjusts to the very recent increases in temperature, sea levels will inevitably rise over time. Studies of the past predict that if we do nothing to lower temperatures, sea levels will rise by 120 feet in coming centuries. How fast this rise happens is anybody's guess, but change in coastlines and the oceans is, to pardon the pun, already baked in.

Another piece of history can be found in octopuses, or more precisely, their genomes. Researchers looked at the genomes of two different subspecies of octopus, one that lives in the Weddell Sea north of West Antarctica and another that lives in the Ross Sea south of West Antarctica today. They found evidence in the gene sequences that the two different populations interbred 100,000 years ago. There is no way they could interbreed now, as they live on opposite sides of West Antarctica. The only way this genetic intermixing could have happened is if there had been no ice barrier in West Antarctica 100,000 years ago. The genes of these octopuses show that what we know today as West Antarctica was indeed completely absent at that time and is therefore as susceptible to climate change as Mercer originally thought.

If the history of rocks, ice, and genomes is any guide, our future will be one of rising seas. While West Antarctica has the potential for the most rapid change under most scenarios, there is even more ice in East Antarctica and Greenland.

Eric Rignot and his colleagues used satellites to measure the height of the ice, its mass, and the movements of ice streams, and then were able to map where and how much ice has been lost across all of Antarctica since 1979. Ice mass loss per year from 1979 to 1990 was about 40 gigatons per year. In the following decade, that number increased to 50 gigatons per year. Then, the rates took off: 166 gigatons from 1999 to 2000, then 252 gigatons per year from 2002 to 2017. The loss happened in both East and West Antarctica, but mostly in specific places in each region.

Rignot's team also focused satellites on Greenland to see how the ice there is holding up. Looking at 206 glaciers in images from 1972 to 2018, they found that Greenland shifted from gaining ice, on the order of 47 gigatons per year from 1972 to 1980, to losing ice every decade afterward—51 gigatons in the 1990s, 187 gigatons in the 2000s, and 286 gigatons in the 2010s. All of this water drained off of Greenland, and this has raised global sea levels by half an inch during this time, mostly in the last eight years. The concern now is obviously not the overall sea-level rise coming from melting glaciers—it's about how fast these changes will happen. With slow changes, humanity can adapt to rising seas. But the rate of ice loss is steadily increasing. Antarctica is now melting about 150 billion tons of ice per year, and Greenland is losing about 270 billion tons per year. And this freshwater all flows into the ocean.

The National Oceanic and Atmospheric Administration (NOAA) gathered data from satellites, work on the ice, and the historical records of sea-level and climatic changes, and then released its 2022 *Sea Level Rise Technical Report*. Sea levels rose 6 inches in the past century. In the next thirty years, the levels are projected to rise nearly 1 foot, rising more on some coasts, less on others. After 2052, the projections become very challenging. The question is not whether sea levels will rise. The issues are: How fast and how much? Putting real numbers on these predictions is challenging because of the complexity of the system and the large amount of unknowns in our understanding of the behavior of glacial ice. As expected in a situation such as this, predictions differ among analyses. One study projects that sea levels could rise as much as 10 feet globally in the next century if the planet warms more than 3 degrees Fahrenheit. But this alarming prediction stands in contrast to a less dramatic one that envisions 4.5 inches of sea-level rise by 2100.

The lives of the kids in that video call linking Grise Fiord and Cape Cod are, like those of the rest of us, bound by solid ice and liquid water. Even modest amounts of rapid sea-level rise can disrupt lives across the globe. Coastal cities and regions that sit close to or below sea level—such as the Netherlands , parts of Denmark, southern Florida, Bangladesh, and other places—will be inundated by saltwater. Saltwater will disrupt geographies, ecosystems, and even drinking water for communities around the world. Pacific island nations such as the Marshall Islands and Kiribati, which sit almost at current sea level, are at risk. The sad irony is that these same islands were regularly decimated during nuclear tests by more developed nations in the 1950s. The industrial activities of many of those same countries today could lead to the disappearance of the islands entirely.

Coastal regions are not the only places that will feel the effects of melting glaciers. The rotation of the Earth and the topography of its coastlines and seafloor drive the ocean currents that move water across the globe. For example, warm ocean waters flow in the Atlantic Ocean, where they cool and then submerge to return south. This current is the reason that much of Europe is warmer than other places at similar latitudes: Madrid and Philadelphia sit at similar latitudes but have very different mean annual temperatures. Melting ice caps in Antarctica and Greenland will push cold freshwater into these currents and are predicted to slow them down. Slowing currents will mean that Europe will become colder and dryer while the southern latitudes get even warmer and wetter. These changes will affect local weather, food production, and economies.

The regions most affected by melting ice are, of course, the Arctic and Antarctic themselves. Here, the relationships among landscapes, peoples, and other living things are all changing.

# CHAPTER 8

# HEATING UP

On slow winter days in Chicago, I hunker down with aerial photos to visualize how our polar expeditions will unfold during the summer. While the photos are mesmerizing and informative today, they have a dark history. During the Cold War, nations sent airplanes over the Arctic with giant cameras to capture giant black-and-white photographs of the ground from 10,000 feet. These photos were taken to help plan flight paths for nuclear missiles. The shortest distance for potential missiles to take between the Soviet Union and the United States, for example, would be over the Canadian Arctic. Since the topography of the land can exert gravitational effects on missile flights, an understanding of Arctic geography became a top priority. The negatives for each picture were a colossal 10 x 10 inches, and the details they recorded were stunning. What's more, these pictures were taken in stereo, so that if you have a special pair of magnifying glasses, you can see the land in 3-D. Sitting with these images, and these glasses, is like a trip to the Arctic while still at home. The images are so sharp that I can often discern regions with ledges good for hunting fossils, patches of gravel on which to set up camp, and relatively safe routes to hike. After spending quiet hours offline with these photos in winter, I feel like I am seeing old friends when I encounter geological features on the ground in summer.

The images taken in the 1960s have lower resolution than those from more recent satellites—NASA's Landsat and the European equivalent, Sentinel-2A. But they differ in another very important way: there is significantly less ice on the recent satellite photographs. We proposed two camps for the 2023 expedition to Ellesmere Island, separated from each other by 10 miles. In the old aerial photos, a medium-sized ice cap lay between the two sites. Recent satellite images show the ice has shrunk by almost half a mile and exposed promising new bedrock between both camps. All over the Arctic, glaciers that sit at the head of river valleys have retreated in the past six decades. There is an unsettling irony that emerges from studying these images: the degradation of Arctic ice makes for more places to hunt for fossils.

The communities that live in the region know these changes intimately. Inuit have lived in balance with the Arctic seasons for more than 4,000 years. During the coming of the Dark Season, when there is full nighttime darkness for weeks to months (depending on latitude), seawater freezes into pack ice and the glaciers on land solidify. Rivers that were raging in July become dry valleys by November. The frozen sea and land become stable for sledging trips until they melt with the coming of the sun in spring and summer. With the arrival of lighter and warmer days, culminating in twenty-four hours of sunlight during summer, water returns as the glaciers and sea ice melt. Inuit hunts, traditions, and lifestyles have been connected to these annual patterns for thousands of years. Every region in Canada's North has different names for the months of the year, but each of them reflects this bond to the natural world. In the Northern Baffin territory (Qikiqtaaluk Region), January is known as Qaummagiaq ("bright days coming back"). Other months are named for the cycles of life of caribou, seals, and birds. Nurrait ("caribou are having babies") reflects the month of June. Ukiuliruut, for October, describes the shift toward the dark part of the year.

But these traditional rhythms are shifting.

The Territory of Nunavut solicited comments from residents on their local environment. The quotes tell their own stories. "It is much warmer now, even in winter. It used to be really cold in January and December but it is not so cold now," said Isaac Kalluk from Resolute Bay.

"It is very obvious that spring is a lot earlier than before. The month of April is one where I can cite an example of the changes, perhaps by many other Inuit. The month is generally used for the Toonik Time spring festivities. It was towards the end of the month when the festivities were held prior to the new century, but these last few years, due to earlier spring, the festivities have to be moved up by two weeks," said Jacapoosie Pete from Iqualuit.

And Simeonie Amagoalik from Pond Inlet observed, "I used to go egg hunting, but now it is too dangerous to travel by ice, so I cannot go to the places that I used to go to. I think it is mainly the ice on the sea that has affected me the most."

An article in the *Robb Report*, the luxury lifestyle magazine, strikes a different chord about these changes. On Bastille Day in 2023 the French icebreaking cruise ship *Le Commandant Charcot* arrived at the North Pole with a band playing "La Marseillaise." The author described passengers, outfitted in matching parkas, idling at the pole: "After an hour of taking selfies and building snow igloos in the icescape, with temperatures in the relatively balmy low 30s, we head back into our heated sanctuary for mulled wine and freshly baked croissants. Mission accomplished. Flags planted. Now, lunch."

The travelers, who paid as much as $90,000 for the seventeen-day excursion, received shipments of caviar, foie gras, and other items from France. After assembling to pose for a drone snapshot, they returned to an Alain Ducasse–managed restaurant in a vessel outfitted with a cocktail bar, juice bar, nine wine cellars, a cigar room, and a cognac lounge. They could then work off the extra calories in a gym, spa, sauna with massage therapists, and heated indoor pool.

In this same place 134 years ago Robert Peary and Matthew Henson trekked to the pole, struggling over 10-foot-high sastrugi to sledge only a few miles per day. In the decades that followed, aircraft were able to make the journey and land with skis on the ice. Later, new icebreaker and propulsive technology enabled the Soviet ship *Arktika* to attain the pole in 1977. With ever-thinning ice, the French tourists on *Le Commandant Charcot* first made the journey in 2022.

The Arctic is warming. We know that this change is part of the regular course of events up north—these regions have heated rapidly in the past well before humans had much impact on the planet. In the mid-nineteenth century, Elisha Kent Kane's sledging teams encountered open water in northern Greenland near where Isaac Hayes saw ice a few years later. Dansgaard, Oeschger, and Heinrich deciphered rapid climate shifts and melting ice in the Arctic thousands of years ago from cores in ice and the seafloor. These facts invite the question: Are the changes we are seeing now just a natural blip, or do they reflect a different kind of trend?

Taking the temperature of the Arctic is a surprisingly complicated affair. Just like a human body—where temperatures readings from under the tongue, on the forehead, or in the ear can be different—readings of the Arctic depend on the location where measurements are made. Land, air, and water each heat and cool at different rates. Consequently, polar regions are bristling with sensors: there are thermometers on the ground, placed on antennae several feet above it, in weather balloons, on buoys in the ocean, and on satellites making infrared measurements. With so many kinds of devices continually monitoring the landscape, computer models are needed to pull all the data together.

Any estimate of climate change in the Arctic depends on the time frame in which it is measured and how the region is defined. The most reliable estimates start after 1979, when satellites were first deployed to scan the planet. Moreover, the Arctic region can be defined by sunlight (the region that experiences one full day of darkness in winter or light in summer), ecological features (where the forest line ends), where permafrost begins, and other factors. If the Arctic is defined by a more northern latitudinal cutoff, then proportionately more ocean is included in the calculation. If you define the boundary as somewhere in a lower latitude, then you are including relatively more land. Because oceans and land hold different amounts of heat, these distinctions are not mere "inside baseball" statistics; they make for different estimates of temperature.

Recognizing these challenges, a Finnish research team pulled all the temperature data collected over the past fifty years. When they corrected for the ways that estimates can vary based on geography and timeframe, the group discovered that the Arctic has warmed four times faster than the rest of the world since 1979. And the warming was not even. Parts of the Barents Sea warmed seven times faster than the rest of Earth.

The heat that is now entering the Arctic is causing the sea ice to melt earlier and to freeze later in the year. Mostly open polar seas are now the name of the game for parts of the summer. Alliterative jargon—Arctic amplification—describes the runaway heating that can happen once melt begins. Ice is reflective and absorbs less solar radiation than the darker layers of land or sea. As ice melts over the sea, less light is reflected by ice and more solar radiation is absorbed by water. When the water warms, more ice melts, setting off a feedback loop wherever ice retreats.

The water under the ice of the Arctic is not only heating because it is absorbing more solar radiation. As the water of the Arctic Ocean interacts with that of the Atlantic, it also comes in contact with Gulf Stream currents and becomes saltier and warmer. NASA calculates from satellite observations that almost 30,000 square miles of sea ice is lost every year. That's the equivalent of losing an area almost the size of the country of Austria. Chemical isotopes and features in the local bedrock give us a deeper historical perspective about the presence of ice. The rates of change we see today exceed anything in the geological record for the past 1,500 years. Unless conditions change, one study predicts that a runaway loop of Arctic amplification will lead to sea ice completely disappearing by early fall as early as 2030 and no later than 2050.

It is a new Arctic on land as well. On June 20, 2021, two European satellites recorded a temperature of 118 degrees Fahrenheit on the ground in the Arctic of Siberia. That same summer, hundreds of wildfires scorched the Siberian countryside and blackened villages with smoke. These so-called zombie fires were the rekindled remains of fires from the previous year. With carbon-rich peat underground, the fires had smoldered for months under the ice. The following month, in July 2021, three successive thunderstorms formed in the Arctic. Typically lacking the heat needed to generate thunderstorms, the Arctic is usually free of them. But since 2010 the incidents of summer lightning there have tripled. In recent years, lightning even has struck within 60 miles of the North Pole.

Hundreds of lives were lost in the attempts to attain the North Pole or to find passages in the ice between the Atlantic and Pacific Oceans. These routes will soon be navigable for large parts of the year. The lack of ice cover means new areas will be open for fishing, mineral extraction, and oil and gas exploration. The U.S. Geological Survey estimates that 30 percent of the world's undiscovered natural gas resources and 13 percent of conventional oil lie in the

Arctic. Lithium to power batteries and rare-earth metals for silicon chips are thought to lie in the seafloor as well. Formerly remote and largely inaccessible places will be open for business. The borders that countries recognize within the Arctic are becoming increasingly important. From small communities to entire nations, humanity's relationship with the landscape is changing. With diminishing ice, the tensions between nations—which have been simmering for decades—are heating up.

In our expedition in 2014, our plane rumbled slowly toward the northernmost extent of land in Canada. The route took us along the coast, with a view of Greenland on the right side of the aircraft and Ellesmere Island below. As we veered to land in a small valley, a tiny island half a square mile wide came into view. Known as Hans Island—or Tartupaluk in Greenlandic, meaning "kidney-shaped"—it sits between Canada's Arctic and that of the Danish Protectorate of Greenland. Visited by Greenlandic hunters, and later Elisha Kent Kane and Charles Hall, this barren speck of land about one-third the size of Central Park in New York became the site of a decades-long geographic struggle between two unlikely opponents.

Neither Canada nor Denmark is known as a particularly hegemonic or warlike state, but by the Law of the Sea, both countries were able to claim Hans Island as their own. The territorial dispute involved military and government officials dropping markers of sovereignty during the summer months. The Canadian land claim was established by officials placing a bottle of Canadian whiskey and planting a Canadian flag in the center of the island. The Danes responded with schnapps and a Danish flag.

The dispute was a real one: while some people jostled with booze and flags, the diplomats and lawyers were undertaking serious work. Claims of land in polar regions, including tiny Hans Island, involve possession of the natural resources on land and below the sea. The two governments resolved their dispute by dividing the island in half, east to west. Signed during the beginning of the Ukraine conflict with Russia in 2022, the deal was meant to show that border conflicts can be settled by peaceful means. A Danish commentator even joked that because Canada now shares a border with a European nation, it can compete in the annual Eurovision Song Contest.

While Hans Island was not the site of a desperate conflict, it reflects an increasingly urgent truth about the polar world. A warming Arctic makes these territories ripe for resource extraction, navigation, and other economic activities. Moreover, the Arctic has some of the most extensive undeclared borders among nation states in the world. Vast areas under the polar ice have yet to be claimed. For the Arctic, what's at stake is not just a dot of land—now we're talking about a country's claims to the entire region under the polar ice cap, including the North Pole itself. And one of the tools countries have used to stake these bold claims is science and the scientific enterprise itself.

## Science and Sovereignty

Wings of a wrecked plane, buffed silver by decades of Arctic gales, lie in pieces at the edge of the dirt airstrip at Resolute Bay. Situated like a billboard at the so-called Gateway to Canada's High Arctic, the wreckage announces the dangers of the place. The Avro Lancaster Mk. X crashed in 1950 when it ran out of fuel on its approach. Remote logistics and changeable weather make flight incidents a regular part of life up north. In 2011, a Boeing 737 carrying community members and the staff of the science station in Resolute Bay

The entrance to the runway at Resolute Bay, the Gateway to the High Arctic

lost visibility in dense fog and then crashed into the hillside, killing fourteen out of the sixteen passengers. For decades, helicopters operating in the early spring have suffered whiteouts and crashes that have killed geologists, ecologists, and polar bear surveyors. In July 2023, my team and I had to turn around on our expedition to the Arctic because the helicopter we were meant to use had crashed earlier that season while carrying another scientific party. Fortunately, nobody was seriously injured, but the craft was totaled, and our expedition was abruptly canceled.

The establishment of that remote airstrip was part of a political dance among polar nations. By the late nineteenth century, polar regions were like a chessboard with flags, human beings, and government offices becoming pieces in an international game of sovereignty. Nations from Australia to the United Kingdom carved up Antarctica like a pie based on longitudinal lines. In 1959, these land claims were pushed aside by the Antarctic Treaty, leaving the entire continent open to scientific research. In the northern polar region, however, no such agreement was reached and the geopolitical situation has remained in flux. With unclaimed lands mostly situated in what is today the Canadian Arctic, the area was long up for grabs. In some places, it still is.

Our relationship with polar landscapes, and with countries' claims of sovereignty over these regions changed with the development of aircraft that could land on rugged airstrips. The biggest breakthrough is seemingly the oddest. Canadian bush pilots realized that if they put giant balloon tires on small propeller planes, they could land in remote locations directly on tundra or gravel. In a headwind, all the plane needed was about 1,000 feet of flat and dry surface. Gravel bars, strips of tundra, and large fields abound in the region; airstrips now lie in every nook and cranny of the Arctic. Trips that would have taken months or years merely decades before could happen in hours or days. My colleagues and I fly in modern versions of these craft. The takeoffs are so short that they feel unnatural; I am tempted to pull on my seat to help the plane ascend. Our landings, like the first one in Greenland, bring a surge of adrenaline and jostled innards. In the 1950s the newly outfitted aircraft were game changers, even more revolutionary than putting wheels on suitcases decades later.

The Cold War fostered a broad U.S. presence in the Arctic to attempt to monitor Soviet activity. A series of early-warning stations were strung east to west across the remote islands of the High Arctic to detect potential missile launches. As the Canadian government became enticed by the opportunities in its northern reaches and concerned that other nations might lay claim to its territory, it encouraged a migration of people to the North. First up were officials such as Mounties and government administrators. Canada established bases for the Royal Canadian Mounted Police (RCMP) as well as post offices in different parts of the region. Even though mail at the RCMP post offices was only delivered once a year, the formation of these stations was a part of their land claim. But use of these regions by Canadian nationals outside of government was mostly lacking.

But scientists soon followed. Scientific research has played an important but very different role in national interest at the poles. In the Antarctic, science became the only legal reason for countries to have a presence on the continent. In the Arctic, the role of science was subtler. As with the Apollo program, the development of science in the Arctic was as much a story of national interest and geopolitics as it was one of curiosity to discover more about the natural world and cosmos.

Capitalizing on new ways of strategically placing people across the Arctic, in 1955 Canada launched a scientific survey to understand the geology of the lands within its most northernmost regions. Conceived and led by Yves Fortier, a Quebecois geologist, the plan was to transport aircraft to Resolute Bay by ship, then deploy small teams of scientists across the region. Teams of two to four scientists, students, and Inuit guides were dropped off for a month or more to survey the rocks throughout the area. Named Operation Franklin, after the leader of the Lost Ship Expedition, the goal was to assess the Arctic's potential for mineral, gas, and oil resources.

Eight years after his crews dispersed across the Canadian Arctic, Fortier published a boxed set of maps and scientific papers that revealed its geology in detail for the first time. The maps unfold as illustrations that have color codes for the types of rocks contained in each area; for example, green for Cretaceous rocks, yellow for Jurassic layers, and purple for Devonian Age ones. The beauty of the maps belies their significance. They are the entrée for scientists like me to study the region. To find a fossil fish that is close to the ancestry of land-living creatures—a fish with arms—we zeroed in on regions colored purple. The rocks in these places are the right age and right type to hold the fossils we were looking for. Fortier's paper, and the ones that followed it, became a road map for scientists of all stripes, from paleontologists to climate scientists, to make discoveries in the Arctic.

In 1957, swept up by the success of Operation Franklin, the Canadian government established permanent scientific infrastructure in the Arctic. Scientific bases in polar regions were hot commodities at the time. Russia's Vostok, the United States' McMurdo, and other countries' outposts on Antarctica also sprung into existence. Resolute Bay became Canada's main hub in the Arctic.

With a scientific base in Resolute Bay, the original mandate broadened to deploy scientists across Canada's North utilizing the techniques developed by Fortier in Operation Franklin. Outfitted with small huts to serve as dormitories and communication stations—and the dirt runway—Resolute Bay became the hub for logistic operations in the area. Still in existence, it supports research on subjects ranging from polar bear ecology to archaeology. For six decades since its founding, this effort has grown to support teams from around the world. My team discovered the fishapod, *Tiktaalik roseae*, with this support.

By the 1960s, with the Arctic now open for scientific research, the stations drew talented and often iconoclastic scientists. One of the trailblazers drawn to this new research was Mary Dawson. A paleontologist specializing in mammalian evolution, Dawson settled in 1962 at Pittsburgh's Carnegie Museum of Natural History as a staff researcher. But being on the curatorial staff, the most prestigious academic line at a museum, was completely out of the question for her. As the director of the museum said to her: "No woman will ever be a curator here."

Standing a little under five-three, Dawson was a force of nature. She was renowned for her hardy approach to challenges in the field.

Once on Ellesmere Island, she and a colleague were hunting for fossils along a rocky hillside set along a broad valley. The two noticed six white dots meandering in their general direction. As the images came closer, it became obvious that these were six Arctic wolves on the hunt. Stalking the two paleontologists, the wolves ultimately encircled them. The alpha wolf charged Dawson, leaping for her face. Dawson stood her ground, spread her arms, and let out a cry at the airborne wolf. The animal dropped to the ground, turned, and walked away from Dawson and her companion, occasionally gazing back at them as it retreated. The other five wolves followed. Some of the alpha wolf's saliva remained on Dawson's cheek.

The robust character that repelled a hungry pack of wolves led Dawson to conquer the world of academia. Ten years after the director at the Carnegie derided the possibility of female curators, she became curator and chair of its Department of Paleontology. Dawson went on to be awarded the Romer–Simpson Medal from the Society of Vertebrate Paleontology, the highest honor in her field.

Mary Dawson's work exemplified the need to take the long view to be successful with Arctic science, something that inspired people like me who followed her. Dawson's early trips offered little in the way of scientific results. But by sticking to her plan over several more expeditions to Ellesmere Island, she and her team discovered 50-million-year-old fossil alligators, turtles, and other warm-adapted species. A previously unknown ancient ecosystem emerged from Arctic rocks—hippo-like creatures, shrews, and other mammals lived alongside diverse reptiles and plants.

The fossils solved an old mystery. Dawson and her colleagues' discovery of fauna that had similarities to North American species as well as to European ones led them to the conclusion that millions of years ago, Europe and North America had been connected by a

Mary Dawson in the field collecting mammal fossils

polar land bridge. Putting geological and paleontological pieces together, she surmised that 50 million years ago, the Arctic was warm and animals could move between the continents more freely. We take continental drift and plate tectonics as truisms today, but in the 1970s, they were still controversial theories in some circles.

The movement of the Earth's plates and the similarities between Arctic fossils with those elsewhere on the globe are not merely of academic interest. They hold a key to understanding the future of the region, even the relationships between countries.

The name of the program that has supported science in Arctic Canada seems out of sync with what it actually does. The Polar Continental Shelf Project (PCSP) wouldn't seem like the obvious name for a government program that supports paleontology, polar bear ecology, lake ecology, archaeology, and fields of basic and applied science. But if anyone seeks to understand national sovereignty, and science's role in it, the moniker explains everything.

In 1948, a Soviet expedition to the Arctic Ocean was doing the routine work of mapping the beds on the ocean floor. As World War II ended, and the Cold War ramped up, an understanding of the bottom of the Arctic Ocean was gaining new significance. The Arctic was close to both North America and Russia, and so the area became an obvious zone of contention. Ever since Nansen's trip on the *Fram*, it was thought that the floor of the Arctic Ocean would be shaped like a bowl, with its greatest depths at its center. But here, the Soviets found something unexpected. While they had envisioned this part of the sea to be over 10,000 feet deep, the center of the ocean extended to a depth of only 3,200 feet.

This is not an easy region of the world to study. Just getting to the spot requires specialized icebreakers. The weather is so hostile and unpredictable that only a limited number of days can be spent working. Even worse is the fact that there are only a few short months at the height of summer with sufficient light. Visualizing the depths of the oceans is not an easy feat anywhere in the world. But in a place where ice covers much of the water, drilling becomes near impossible. A team is lucky if they have enough time to make a few soundings of the seabed to assess depth and dredge the bottom to collect samples from the floor.

Despite these challenges, another Soviet expedition returned to the area in 1954. Tracing the ocean floor, the team discovered that the shallow area was the summit of a previously hidden undersea mountain range. While they were not able to trace the mountain's full extent, they theorized that the range extended as a ridge across the length of the Arctic Ocean. They named the feature the Lomonosov Ridge, after Mikhail Lomonosov, a poet and natural philosopher who is thought to have predicted this structure in the 1800s.

More measurements confirmed the Soviet team's original hypothesis: the Arctic Ocean is not composed of a single basin; it is made of at least two of them, divided by the 1,100-mile-long Lomonosov Ridge. Rising almost two miles from the ocean floor, it is a natural barrier that affects ocean currents, marine life, and ice activity. But what makes this geological feature so important for the coming decades is where it is situated. The ridge is connected to Russia's Siberian islands on one side and extends underneath the North Pole all the way to northern Greenland and Ellesmere Island. Mary Dawson discovered an ancient connection between the Arctic and Europe. Here, underneath the sea, is one that exists today.

This is where the name of Canada's research enterprise—Polar Continental Shelf Project—gains real meaning. Science and national sovereignty are intertwined by the Law of the Sea. By international law, nations are entitled to the resources based on the position of their continental shelf. Continental margins extend under the sea as a shallow shelf before diving to the depths of the ocean. Countries are entitled to exclusive rights to the seafloor 200 miles beyond this feature. If a nation can show that a shelf is part of its continent, that country then owns the seafloor and everything inside of it—all the minerals, oil, and natural gas. The waters would remain open for navigation and fishing, but the extraction of resources would be restricted. Since the Lomonosov Ridge cuts the Arctic in half under the North Pole, it raises the stakes for national claims. What's up for claim is hundreds of thousands of square miles, an area significantly larger than the state of Texas. The Lomonosov Ridge is central casting for a geological structure that can cause international disputes.

In the summer of 2007, three Russian scientists descended in a mini submersible 12,000 feet in the ocean below the North Pole. Once at depth, they extended a robotic arm toward the seafloor. Manipulating joysticks from inside the craft, they wedged a titanium Russian flag into the rocky bottom. Russia was declaring the Lomonosov Ridge part of its own continental shelf and thereby staking claim to the vast seafloor. Once on the surface, the team received a congratulatory call from Vladimir Putin, lauding the nation's extension to the polar north.

Evaluating this claim means knowing the geological origins of the Lomonosov Ridge. There are different kinds of undersea mountains, and not all of them were originally part of continents. The seafloor is constantly spreading and expanding in a process that leaves ridges in the middle of the ocean. Because these kinds of ridges were never on land as part of a continent, they cannot be used to support a claim of sovereignty. The rocks must be shown to have been a part of a continent—and not just any continent, but the one of the country making the claim.

The three nations that border the Lomonosov Ridge—Canada, Denmark (via Greenland), and Russia—have been mapping the seafloor and dredging up rocks to make a claim on the territory. For example, in 2005, a joint Swedish-Danish team traveled by icebreaker and dredged up debris from the seafloor at the ridge. Once they brought rock to the surface, they applied chemical and geological analysis to determine where it came from. The first issue is whether the rock is from the ridge or was simply dropped there by the action of glaciers on the surface. Over time, glaciers grow and shrink and carry rocks in their wake. The boulder the team dredged up had a rind that covered it in layers. Chemical analysis revealed that the rind was magnesium oxide. This mineral is characteristic of nodules that have sat on the bottom of the seafloor undisturbed for thousands of years. The presence of this crust is evidence against the boulder being dropped by a melting glacier because it is too old and shows no signs of movement. If it was not dropped by a glacier, then the boulder was likely part of the ridge itself. When the team looked to the interior of the rock, an ancient geological story emerged.

The inside was composed of folded layers containing the mineral mica, telltale features that the sediments were compressed during the emergence of a mountain range. Moreover, atoms inside the rock told the geologists that it formed 470 million years ago during a well-known period of mountain-building in Greenland. The conclusion was obvious: this rock was part of the Lomonosov Ridge, and its geological origins were part of those that happened in northern Greenland. This analysis was one line of evidence Denmark used in 2014 to propose that the Lomonosov Ridge is part of Greenland and claim 360,000 square miles of the polar region, including the North Pole and part of what is now close to the Russian Arctic.

Research by the Danes and Swedes, as well by as other teams, makes it clear that the Lomonosov Ridge is a strip of continent. And using geological approaches, Denmark, Canada, and Russia can each argue that its composition is similar to the crust on their territories—Siberia on one side and Ellesmere Island and Greenland on the other. The reason is as straightforward as Mary Dawson's fossils. The origin of the ridge is still under investigation, but it is likely that, at some point, it peeled off what is today Russia, Canada, and Greenland as they moved apart in the distant past.

Russia, Denmark, and Canada have each filed claims to the United Nations Committee on the Law of the Sea (UNCLOS) for hundreds of thousands of square miles of territory. The Danes claim that 360,000 square miles is their territory, while Russia and Canada each make a case for 460,000 square miles of their own. Thousands of pages of text, diagrams, and maps have been filed with each submission. The seafloor underneath the Arctic Ocean, as well as the North Pole itself, is now a quilt overlapping claims.

UNCLOS reviews the scientific basis for each nation's proposal.

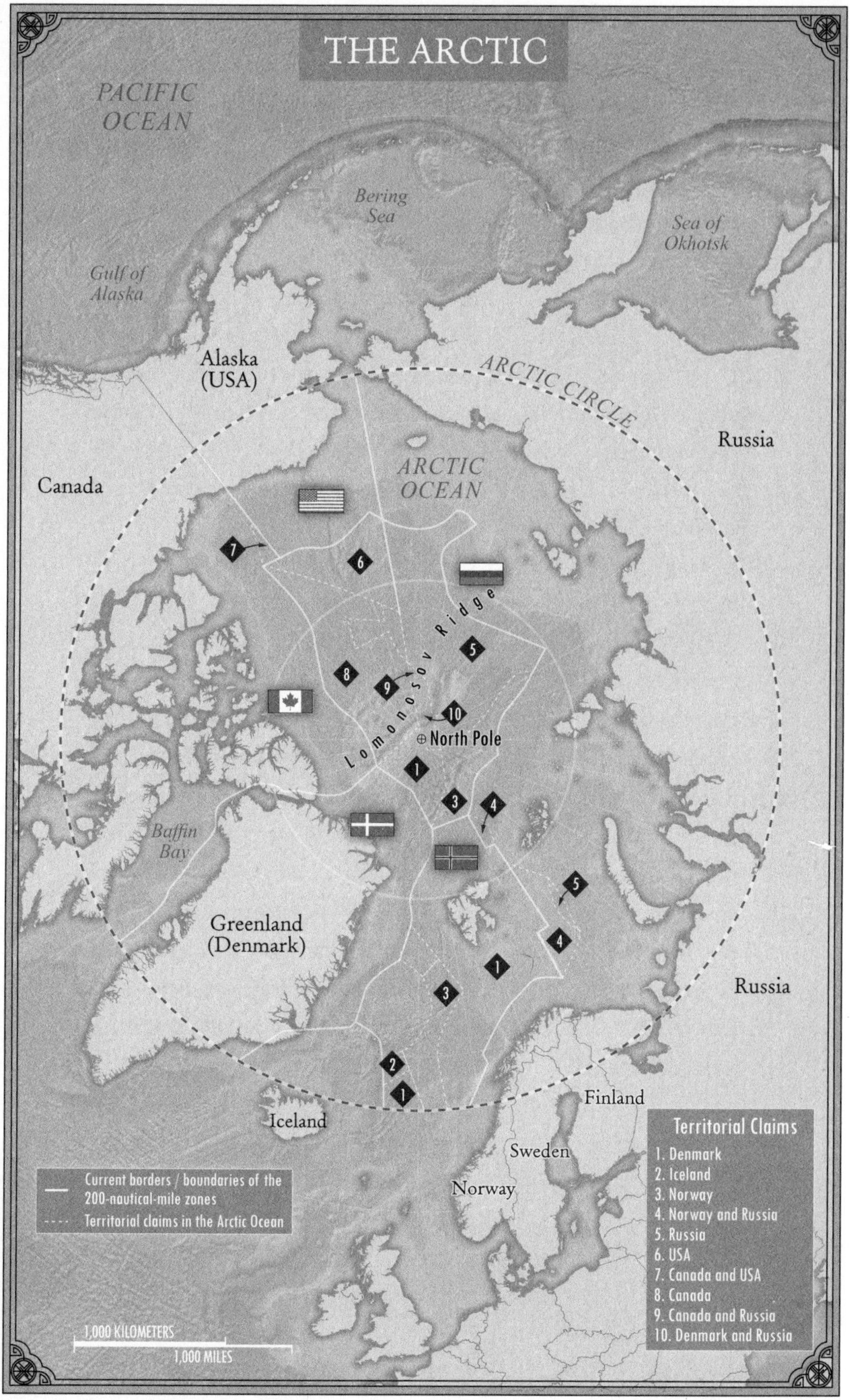


The Lomonosov Ridge and the thicket of claims to polar territory

They ask: What is the evidence that a particular feature is continental shelf, and how far can a country claim exclusive rights? But the committee's job stops there. UNCLOS makes no further recommendation on which country owns a patch of land. That outcome is the job of diplomacy.

Russia submitted its declaration first in 2005 and amended it several times afterward. In 2023, UNCLOS announced that it largely agreed with the Russian claim. They are still evaluating the Danish and Canadian claims in a process that could take a decade or more. It is entirely possible that UNCLOS will agree with all three countries' proposals. If that's the case, then the diplomats will have their work cut out for them.

Territorial disagreements surrounding the Lomonosov Ridge are only a part of the complexity the polar world. The United States and Canada share continental shelves in the northern extent of Alaska and the Yukon Territory. These regions overlap for thousands of square miles. There are also struggles over ownership of sea routes. Can ships of various nations pass through the waters around the islands of Canada's High Arctic as they make their way between the Atlantic and Pacific Oceans? Canada and the United States disagree on this issue. What about a more northern route that connects the oceans along the coast of northern Russia?

The United States may be at a disadvantage in negotiations about northern sovereignty and access. It never signed on to be a member of UNCLOS because of concerns voiced by treaty-adverse senators. In addition, China is now claiming status as a "Near Arctic State" with significant interests in the region and could profit by using Russia's sea routes north of Siberia. A warming Arctic makes each of these routes feasible, leading to disagreements that could not have been foreseen decades ago.

In 2008, Denmark gathered the five nations with claims to the seas in the North—the Russian Federation, Denmark, Canada, the United States, and Norway—to sign a pact of cooperation, the Ilulissat Declaration. The goal was to bring the nations to work together on "vulnerable ecosystems, the livelihoods of local inhabitants and Indigenous Communities, and the potential exploitation of natural resources." It also committed each nation to respect the Law of the Sea and hew to a peaceful resolution of territorial claims.

The ideals of the document are noble and offer hope for the future. But what's missing from the Ilulissat Declaration is important. Arctic nations such as Iceland, Sweden, and Finland are notably absent from the process. There have been calls to use the Antarctic Treaty as the exemplar for international collaboration in the use of the seafloor below the North Pole. This movement seeks to guarantee that no country claims the polar seafloor, and that all nations dedicate its use for science. But the Ilulissat Declaration, in committing only five nations to abide by the Law of the Sea, specifically obviates that possibility. The Antarctic Treaty involved nations negotiating over an uninhabited and mostly unexplored area. The Arctic, home to multiple countries and local populations, is entirely different. And because of that fact, it will be increasingly contested in the coming years.

For over a century, there has been faith in a romantic notion called "Arctic Exceptionalism," the view that the polar landscape fosters collaboration as a means to solve problems and adapt to change. With a certain lens, the history of science, exploration, and international cooperation can be seen in this way. Changes emanating from the south, such as strains between Russia and the other Arctic states, put this ideal to the test. The Arctic Council is a collaborative body that includes all Arctic nations as well as representatives from Indigenous Communities. As of June 2022, the council has effectively estranged Russia from its activities. It is hard to imagine effective policymaking when the single largest Arctic stakeholder is not part of the discussion. While strained diplomatic relations in the south inevitably spill over to the north, the obvious corollary is that there will no longer be consultation on the environment, science, safety, and the needs of Indigenous Communities.

While countries draw new maps in the Arctic, both in reality and metaphorically, the individuals who have had the longest relationship with the land have been caught in the middle. These communities, too, have been tied to polar land claims.

Oral histories, archaeological records, and evidence from DNA show humans first inhabited the Arctic region by 4,000 years ago. Presumably after migrating over the Bering Strait from Asia, this population, known as the Dorset people, spread widely across what is today Alaska and northern Canada. Then, about 2,000 years ago, the Dorset mysteriously disappeared, vanishing without a trace to be ultimately replaced by other migrants, such as Thule. These populations lived in small, nomadic groups that functioned without hierarchy or tribal chiefs.

With the arrival of Europeans, and encroachment on lands, Inuit began to assert land claims and organize. In 1969, the Canadian government proposed to abolish all Inuit treaties' land claims. The ensuing outcry, activism, and litigation ultimately landed in the Canadian Supreme Court, which ruled in 1973 that Inuit land claims are legitimate. This decision led to negotiations among Inuit communities, the Canadian federal government, and other bodies where land claims were formally established. Four regions were ultimately recognized, with the largest one, Nunavut, gaining its own legislature and becoming Canada's newest federal unit in 1999.

The road to this achievement was anything but smooth. Disruption has long been part of life for Indigenous Communities across the Arctic.

## Displaced

"We were on what we called Prison Island." This was how Larry Audlaluk, now an elder in Grise Fiord, described his childhood home to a journalist in 2017. It is hard to imagine Ellesmere Island as anything but a pristine, wild landscape where both humans and animals run free. But to eighty-seven Inuit, it became a remote place to hold people. There, Audlaluk and his community became, in his words, "human flagpoles" in a geopolitical struggle for control of the Arctic. Canada was relocating Inuit communities in order to lay claim to the northern islands.

The Canadian Coast Guard ship *C. D. Howe* was designed to be a visible presence in the Arctic islands. Reinforced to travel through ice, it conveyed supplies and goods to remote RCMP stations and weather stations throughout the Arctic. Starting in 1953—two years before Operation Franklin—the *C. D. Howe* gained a new role. It now was going to bring people north to reinforce national sovereignty of the Arctic.

When Audlaluk was two years old, the RCMP lured families from Inukjuak in Northern Quebec to Ellesmere and Cornwallis Islands with enticing promises. Offered a land of richer hunts and hospitable climes, Inuit were assured of a new and easier life than the one they were experiencing in Quebec. And as was done in many a land deal, the families were offered a full refund if the new conditions did not please them. Should the new locale not be to their liking, they would be returned home after two years at no cost to them. The only caveat was that the families could bring only a small portion of their possessions on the new journey.

The *C. D. Howe* conveyed the families north, stopping in the existing High Arctic town of Pond Inlet to pick up people who knew how to live in the region. These residents were to serve as mentors for incoming families. Upon their arrival, the deal started to look dicey. As the travelers approached the Arctic islands, related clans were told that they would be split up among three locations, one in southern Ellesmere, the other in Resolute Bay on Cornwallis Island, and another at Alexandra Fiord. Distraught at being separated from family members, they now feared that there was a hidden side to the deal they were promised.

Getting off the ship in their new home, the arrivals were stunned by the barren landscape of rock, mud, and ice on southern Ellesmere and Cornwallis Islands. Their familiar world of tundra was gone. The situation became more challenging as the year progressed. Here, unlike Quebec, the land is dark for six months every year. Audlaluk, writing in his 2021 memoir, *What I Remember, What I Know: Life of an Arctic Exile*, described how his mother, not used to the dark, went to chip ice to melt for drinking water. She collected rocks instead. Blackberries, rhubarb, and other plants that had formed their diet farther south were absent. Their houses of sod leaked and did not offer the same comforts and protection of their housing back home. Wildlife restrictions imposed by the government controlled where and when the families could hunt. Limitations in diet and housing, the continual darkness of the winter months, and displacement from their traditional relationship with the land took a toll on their health.

Audlaluk had been a toddler when he moved from Quebec to what would later become Grise Fiord. He had the resilience of a child. But his parents were a different story altogether. Upon arrival on Ellesmere, Audlaluk's father became quiet, sullen, and reserved. Being displaced from his native geography, he made his sole focus to return home to Quebec. Suffering fainting spells, he died suddenly only a few months after moving, while the family was inside the shelter cooking fish.

The nearest place to trade was a full day's journey from the com-

Larry Audlaluk

munity. The RCMP station at nearby Craig Harbor was resupplied with goods and staples by the *C. D. Howe*. But the community had only sporadic access to sugar, tea, and flour. This scarcity was by design. Inuit were kept away from the stores for fear that they would get acclimated to handouts and not develop their own independence. This restriction made for famine during regular intervals of the year, particularly February through May. The memory of these deprivations persists, and almost fifty years later, Audlaluk says, "Until very recently, I have had to struggle to enjoy the spring."

The assessment of the situation by the RCMP was entirely different from Inuit experience. Writing for the *RCMP Quarterly*, a Constable Fryer described how initially Inuit were "a depressed, lifeless group of individuals, who were looking for too many handouts from the white man." However, after a few short months he reported that they had gained vigor and "looked so much more alive" in their new location. Judging from Audlaluk's account, this report was untethered from reality.

Removed from their traditional relationship with the land in the subarctic world of Northern Quebec, Audlaluk's community had to adjust to the strange new cycles of light, ice, and plant and animal abundance of the High Arctic. Audlaluk's family was moved from their land so that Canada could claim other lands as their own. The Qikiqtani Truth Commission was founded in 2000 to probe the events that happened between 1955 and 1975 in these territories. Responding to the commission's reports, the Canadian government offered a formal apology in 2019 to those they'd relocated. But the work of reparation is still under way. For example, the historical displacement of Inuit families continues to have ramifications. Life in Grise Fiord is extraordinarily expensive—a trip to the local coop for canned goods can cause sticker shock: prices are several times higher than those in the south. Without support to the communities, one of the legacies of the disruptions of the 1950s will be continued economic struggle.

Members of the Yup'ik Indigenous Group from western Alaska have encountered a different kind of displacement. Living in a subarctic tundra ecosystem, their traditional lands cover wetlands and flat tundra adjacent to water—either braided rivers or the coastline of the Bering or Chukchi Seas. They were traditionally seminomadic and moved seasonally throughout the year, with summer camps in territories that had access to fish and berries, and winter ones close to hunting regions for mammals. When the Bureau of Indian Affairs of the United States began building schools in the late 1950s, life changed. The fraught history of these schools is derived from the trauma experienced by students—many of whom were forcibly relocated from their families—and entire communities. In the village of Newtok on the Ningliq River about 500 miles from Anchorage, for example, a school was built near the community's winter settling place, across the river from where they fished in summer. As the new school developed—with the federal requirement that Indigenous children attend—a permanent community with housing, an airstrip, and a tribal center formed around it. This shift from a traditional semi-nomadic way of life to a fixed settlement has had profound effects on their lives.

## A New Landscape

The first settlers of Newtok, Alaska, had to squint to see the rivers in the distance. With melt every summer, the adjacent river started to get closer and closer to the village. Concerned about the pace of the river's erosion, the village received funding in 1983 to assess the changes happening to the land around the village. Comparison of aerial photos taken over the course of thirty years revealed that the bank around town was losing between 18 and 88 feet *per year.* In 2017, Newtok elder George Carl told the Wilson Center in Washington, D.C. (translated from the original Yup'ik), "In my younger years, where I stood in the community, I would look miles and miles away, but never see the river's edge. Today, if you look out from my home, the edge is really just a few steps away."

At that rate of erosion, the village—including the school that anchored the formerly nomadic community in place—would be lost to the river waters in a few decades. With more and more of the village being engulfed by water with each passing year, it became clear that a move was necessary. Lobbying the Federal Emergency Management Agency (FEMA), the residents and Alaskan congressional delegation were able to secure federal funds to begin to assemble a new village in a stable spot 9 miles away.

Moving a village is profoundly complicated, particularly in a remote location. Managing a retreat is as much about retaining culture and values as it is about providing resources, logistics, and funding. Some people do not want to move from their homes. Others resist moving to a community that is only partially established. In this remote location, getting the supplies and people to do the construction is limited. With housing still insufficient to accommodate all the people of Newtok in the new location, the population remains split between two villages, with neither one fully operational. As of 2023, only seventy people have moved, leaving nearly two hundred at the old site.

In November 2022, the federal government committed $50 million to relocate two Alaskan villages, Newtok and Napakiak. While Newtok has lost an average of 70 feet of land per year in recent decades, Napakiak, about 100 miles to the east, on the Kuskokwim River, is losing 25 to 50 feet of land per year. These two communities are test cases for the process of relocating. Other villages will soon have the same needs. In 2019, the Denali Commission found that 29 Alaskan communities were at risk for erosion, an additional 35 for permafrost thaw, and another 38 for flooding. And those figures are only for the Arctic of the United States. More communities in Canada, Siberia, and Greenland are at risk. As George Carl of Newtok told the Wilson Center (translated from the original Yup'ik), "In the past, when I was younger, I was told about global warming events that might happen. Back then, I didn't realize what they were talking about. But now, with what I see in my community today—there is very little snowfall, water no longer ices over, everything is melting."

At the center of these changes is permafrost. It is deceptive to encounter permafrost for the first time. The surface of the tundra can be soft, squishy, and wet. It might even have muds that suck legs down to the knee. That area is appropriately named the active zone because it freezes and thaws with the passing of the seasons. With increasing depth, layers become frozen as hard as concrete. We've broken tools and tent stakes in places where the frozen soil and rock are close to the surface. The subterranean ice, often embedded with ancient tundra, peat, debris, and organic matter, can be as much as 5,000 feet thick in Siberia. In Alaska, it can be 2,000 feet thick and gets thinner toward the south.

Permafrost, a region where ice persists underground for at least two years, makes up about 15 percent, or 6.9 million square miles, of the Northern Hemisphere. Along with the ice at the poles, it is the planet's freezer. When glaciers extended from the poles to lower latitudes, the earth beneath them froze. When the ice retreated 12,000 years ago, it left frozen earth in its wake in higher-latitude locations. Trapped inside the ice below ground is the world's biomass from long ago—soil, gravel, mammoths, reindeer, caribou, and the body of the occasional human, as well as plants, fungi, and microbes.

Seen from a plane, the surface of the tundra reveals a world of different shapes—hummocks, cracks, elevated polygons, and craters. It is easy to become spellbound by the sheer variety of forms. Melting permafrost changes the landscape because liquid water takes up less volume than ice. When ice thaws, the ground above buckles and a crater can appear at the surface. In some places, the tundra looks like a green version of the surface of the moon. At Batagay, in Russia's far east, a crater started to form as a small ravine in the 1970s. Now it is a gash in the land almost a half a mile long that attracts climbers and cavers. The crater is expanding by several feet per year, perhaps as much as 20. In the summer of 2020, the land below a large diesel refinery in the Russian city of Norilsk collapsed, causing a spill so large that it has been compared to that of the *Exxon Valdez*. The land under Nunavut's capital city, Iqaluit, is buckling so much that water, heating, and sewage pipes need to be continually replaced.

The erosion of the riverbanks by Newtok and other Alaskan villages is accelerated by warming. The banks of the rivers were formerly stabilized by hard permafrost. As the permafrost thaws, and the flow of the rivers increases, the banks collapse. For a long time, it was thought that permafrost thaw and erosion were a gradual affair. At Newtok, when a storm hits and the flow of the river surges, almost 70 feet of land can disappear in a matter of days. Elsewhere, the changes can be even more rapid. In 2015, in Canada's Northwest Territories, a cliff bordering a tundra lake collapsed as it thawed. An estimated 800,000 gallons of water escaped into the adjacent Peel River in two hours. Silt, peat, and heavy metals in the permafrost flowed into the river, leaving a trail that extended for miles.

Like those of ice itself, the dynamics of permafrost melt are counterintuitive. The more snow cover there is over the surface, the faster the melt of the permafrost. The reason is that, like an igloo, snow forms a barrier for insulation, keeping the ground below warmer than it would be otherwise. In areas with more snow cover, permafrost is at greater risk of thaw.

Communities in the Arctic are developing ways to adapt to these changes. Reinforcing buildings with struts and pylons dug into bedrock stabilizes them. Building houses on stilts, or elevating them with chocks, keeps them off the surface of the permafrost. These structures retain heat that could melt the ice below. Communities are now experimenting with pumps that remove the heat from the permafrost and delay their melting.

Melting permafrost releases organisms and remains that have been trapped for millennia. When organic matter thaws from the ice, microbial action generates methane and other greenhouse gases, which ultimately get pushed into the atmosphere. This activity is a source of carbon that can rival the burning of fossil fuels by humans. Within the permafrost is one of the largest carbon reservoirs on the planet. By some estimates there is more than twice the carbon locked within permafrost than there is in the atmosphere. Changing permafrost sets off another feedback loop—melting ice and permafrost release carbon into the atmosphere, which, in turn, causes more warming. With more warming comes more melt and more carbon in the atmosphere. And so on. This is a cycle that, once started, will take decades to slow down.

People on the Marshall Islands, Kiribati, Tuvalu, and other remote, low-lying Pacific islands are relocating to higher ground because of rising sea levels. At the same time, Arctic communities such as Newtok are also moving to greater elevations because of thawing ice and eroding land. Ultimately, melting land and rising seas will affect us all.

## Survival of the Resilient

New patterns of sea ice and permafrost affect the animals that inhabit these regions as well. The ebb and flow of light and dark, water and ice, are part of the cycles of life for all living things in polar regions. Animal breeding patterns, migrations, and growth spurts are timed to annual rhythms. The life cycles of everything from large polar bears to small woolly bear caterpillars are set to rotations between light and dark, warm and cold, calm and windy. Changes to these cycles affect animals and plants as well as the human populations that depend on them.

Polar bears are the most iconic Arctic creatures and conjure wonder, awe, and fear. They have become potent symbols of climate change—with the declining sea ice, their habitat, food sources, and ability to reproduce are threatened. Given these facts, and ever-increasing rates of Arctic warming, some population models predict their extinction by the year 2100. While these scenarios may ultimately be proved true, there is much that we do not understand about polar bear populations and how they respond to changing environments.

Scientists have identified 19 polar bear subpopulations spread across the Arctic, from Alaska to Siberia. Taking a census of polar bears involves counting their numbers from the air, recording their health and vitality on the ground, and assessing the number of males, females, and offspring in a location in any given year. It is not possible to do these kinds of measurements across the whole of the Arctic, and some regions remain better sampled than others. Moreover, historical conjectures about population sizes that were made decades ago are not reliable because we have no rigorous assessments of populations from the 1950s and 1960s. Using more recently collected data, researchers have observed that of the 19 subpopulations of polar bears, 4 are seeing declines in population numbers, 5 are stable, and 2 are increasing their numbers of bears. Eight additional subpopulations are so poorly sampled that we do not know how the animals are faring.

With the gaps in our knowledge, and the different ways in which the subpopulations are changing, it is difficult to come up with a global assessment of the health of the species. Looking at a single region from year to year may be the most rigorous way forward. The areas that are undergoing the biggest population declines are mostly in the southern regions of the Arctic that are also witnessing the greatest reductions of ice cover. In the southern Beaufort Sea subpopulation, for example, polar bear numbers are rapidly declining because the loss of sea ice leads to shortages of food. Polar bears in other areas are responding differently. Populations on the Norwegian island of Svalbard have remained relatively constant, despite significant degrees of warming over recent decades. There may be local factors at work that dictate a population's response to climate change. These details matter if we are to conserve the species.

In 2004, an Icelandic geologist working on Svalbard found a fossil polar bear jaw in the rocks. This discovery was a lucky break.

Since polar bears mostly live on ice, not on land, their carcasses often end up in the ocean and do not make it into the fossil record. Given the rarity of the find, researchers studied it with cutting-edge tools to extract as much information as possible: they looked at the atoms inside the bone, the molecular biology of its tissues, and its anatomy. The specimen turned out to be an adult male that lived between 130,000 and 110,000 years ago. The age of the jaw leads us to an important question about conservation. The era when this male polar bear roamed Svalbard was warm, even more so than today. If the species survived during a period of low ice, can it do so today?

Hints of answers come from the genome. The international team of scientists was able to extract high-quality ancient DNA from the jaw and compare it with that from current populations of brown, black, and polar bears. The DNA was a trove of data. The three species of bear have been largely distinct for 1.6 to 1.3 million years. During this time, brown bears and polar bears occasionally interbred. We tend to think of species as populations that cannot reproduce with one another, but brown bear and polar bear populations share 5 to 10 percent of their genes. The two species aren't entirely genetically separate—the distinctions between them are "messy," according to the study's authors.

Polar bears survived warm periods in the past and have showed resilience in the face of climate change. The scientists hypothesize that during times of diminished ice over the past 500,000 years, polar bears found refuges on land and occasionally interbred with brown bear populations that ventured farther north in a warming climate. The input of the brown bear genes may have helped the polar bears adapt to the warmer climates.

But times are different now. The DNA analysis shows that polar bears alive today have less genetic diversity than their ancestors. This change inside the genome means that the bears may have fewer genetic resources to enable them to respond to environmental fluctuations. Modern polar bears may have limited ways to adapt in the future.

Warming polar regions bring species that have adapted to lower latitudes into the higher ones. These new arrivals can trigger changes across the ecosystem. Consider the orca. As temperatures rise in Arctic seas, orcas extend ever northward. Narwhals, the "unicorns of the sea," have evolved a tooth into a single sharp tusk. Lacking a dorsal fin that could otherwise serve as a big rudder, they have also specialized to live right under the ice. But with receding ice and increasing numbers of orcas moving into Arctic seas, narwhals are becoming ever more vulnerable to predation. By some estimates, 1,500 narwhals are now lost to orcas each year. The same is true for beluga and bowhead whales. Bowhead whales used to live entirely free of predators. Orcas are now affecting their populations too. Scientists estimate that orcas will soon be the Arctic's top predator, replacing the polar bear. There are also changes on land. Red foxes, for example, are a relatively large southern species that is now extending its range up north. They are replacing the smaller Arctic fox that was native to the region.

We are entering a world of uncertainty: changing interactions, habitats, and species ranges will affect food webs, individual species, and whole ecosystems. Grise Fiord and other northern communities are having to adapt, yet again, to new cycles of the seasons, ways of trekking through the ice to traditional hunting locations, and different numbers of whales. As orcas become ever more prevalent, the traditional hunts for narwhals, beluga, and bowhead whales are at risk. For Inuit, narwhal hunts are not only about food, they are also about culture, intergenerational tradition, and strengthening community bonds. When a hunter kills a narwhal, they typically do not salt it and save the meat for later; they share it with neighbors. A hunter from northern Baffin Island, James Simonee, told a journalist in 2022, "There are a lot less narwhal. Two years ago, I caught one. But I haven't got any since." Diminishing hunts affect not only a community's nutrition, they alter the social glue that has been around for generations.

Changes in the seasons are creating new opportunities for some species. The woolly bear caterpillar used to have only a few weeks to go through its annual life cycle of feeding and hibernating until reproducing. With summers lengthening, it is expected that their numbers will rise. Resilient species, the ones lucky enough to be able to shift their breeding cycles, food sources, and habitats, will flourish. Species confined to very specific polar habitats and food sources, or those with low genetic diversity, are on the margins and will go extinct unless the situation changes. Arctic fox are a threatened species because they are specialized for habitats that are shrinking. This species is also coping with the entry of the larger red fox in their range.

The name of the game for survival in polar regions, whether for humans or other living things, is adaptation. Darwinian evolution has been caricaturized as "Survival of the Fittest." While that phrase is an oversimplification of the evolutionary process, it is fodder for a metaphor of life at high latitudes. Success and longevity in the polar world is a story of Survival of the Resilient.

# EPILOGUE

Moving ice has transformed our world. *Homo sapiens* arose during a rare moment in our planet's history, one with ice covering the polar regions. Ever since, we have lived with land, coastlines, islands, and ocean levels defined by the amount of water locked in polar ice. But we have become increasingly reliant on the health of these regions just at a time when our activities are rapidly altering them.

These changes force some scientists and engineers to ask new questions: Can we use our knowledge of polar glaciers to make them more resilient to global warming? What if we could engineer glaciers to resist melting into the sea? There are conceptual blueprints to construct deep-sea barriers to prevent water from flowing underneath coastal glaciers such as Thwaites and Pine Island. These dams could keep warm water from melting glaciers and could halt their retreat. One way to achieve this feat would be to build a rock berm underneath the glacier or build an artificial island to stop the glaciers' flow. Still other ideas envision deep boreholes from the surface to the base of the glacier to extract meltwater or to inject freezing water to keep the base cool.

The cost and ambition of these ideas seem almost impossibly high. But engineering scenarios become more realistic when we project that melting glaciers could lead to more than 2 feet of sea-level rise by 2100. This change would result in the displacement of billions of people, with the global cost of the damages estimated to be as high as $50 trillion per year.

To some, thinking about glacial engineering is utter hubris—humans would be fiddling with entities they do not completely understand and could set off a cascade of unknown consequences. To others, this approach is an insurance policy in case the situation with polar melting gets out of control. And to still others, this thinking is a necessary outcome of the changes we have already set in motion by not controlling carbon emissions.

Regardless of one's take on geoengineering, we have moved from seeing polar worlds as vast, wild landscapes to places that can be directly manipulated, even possibly designed.

Human activities have touched every corner of our globe. With ever-expanding satellite connections, my students can now stand with their lips pursed taking selfies in front of glaciers that only sixty years ago were mostly inaccessible. While it is easy to reminisce about the good old days of remote and wild landscapes, these changes lay bare our real connections to polar worlds. We have been dependent on polar regions all along in ways that have only increased in the past century and a half. The ways we live in lower latitudes affect the ice in the Arctic and Antarctica, which, in turn, influences our own environment.

Living and working in polar landscapes is as much about values as it is learning to survive in harsh environments or making scientific observations. Insights can emerge from unexpected encounters with the terrain.

I nearly wore out the soles of my boots at our first Greenland camp in 1988. Farish had scrutinized crude aerial photos and maps to find a promising campsite before the expedition. Once we hit the ground, however, it became clear that we were a long hike from any rocks that could hold fossils. The trip to good outcrops took us miles along a valley floor through fields of eroded boulders, pebbles, and silt that forms muds during the early season snowmelt. Fossils were the last thing on my mind on hikes. I walked in zigzags to avoid viscous muds while trying to maintain my concentration in the din of incessant polar winds.

One day, my attention was jolted away from the rocks, mud, and wind by the sight of an injured shorebird with an orange beak and black feathery ring around its neck. The bird seemed to arrive out of nowhere; it was as if it had magically appeared 10 feet ahead of me. It was clearly in distress, and its urgent cries rose above the roar of wind as it limped along, dragging one wing against the ground. Its pain triggered an instinctive need to help. As I approached the bird, its wings fluttered and it flew 6 feet farther away from me. Then the bird resumed its anguished call and limp. I approached again, but the bird's wings flickered as it flew another 6 feet away. Not wanting to add to the creature's discomfort, I reluctantly resumed my trek to the Triassic rocks ahead.

That evening in the kitchen tent, Bill and I dug though my field guides to get the lowdown on the bird. Ringed plovers lay their eggs directly on Arctic ground and need to protect them from roving foxes, owls, and raptors. Plovers have evolved a clever strategy to deceive intruders. As predators approach, the birds feign injury with an anguished cry and a fake damaged wing. They use this dance as a decoy to lure threatening animals ever farther away from the nest. Like a sucker, I fell for a classic avian ruse.

Passing by the site over the next five days, I kept an eye out for the nest while trying not to disturb the parental birds. After a few tries, I happened on three gray eggs, each about the size of a large Brazil nut, lying in a small depression in the rocks. Other than a few stones splayed apart as a cradle under the eggs, the place was indistinguishable from the rest of the rocky landscape. There was no cover, no wall around the nest, and no soft bedding supporting the small eggs. The eggs sat completely exposed to polar storms, gales, and predators—their only defense being camouflage and an adult that can fake an injury. Plover eggs embody the fragility, resilience, and endurance of all polar life: for thousands of years plover chicks have hatched while lying virtually naked in one of the most forbidding environments on Earth.

Through more than three decades in polar regions, that image of plover eggs has sat inside me, whether on traverse or in a helicopter over glaciers. Journeys to the ends of the earth help us see ourselves and our home in new ways. Like the plover, humans live in a fragile balance with the planet, one that—for those of us who do not live at the poles—is masked by our built environment. A human body is a speck on a continent-sized sheet of ice. Our lifetimes are tiny moments in the multibillion-year history of the meteorites inside glaciers and the rocks in valley walls. And we live in a relationship with a planet that is capable of dramatic environmental change.

But rather than make us feel small and insignificant, polar regions can enlarge us. Scientific discovery at the poles reveals our deep connections to one another and to the planet. Here, life is only possible when humans look beyond themselves to survive, thrive, and adapt. Our existence on frozen tundra and ice is impossible if we do not learn from and collaborate with one another. And the more we look, the more we find that polar regions are themselves fragile places, ones that can change in a brief moment of geological time. It is profoundly humbling to find our own vulnerabilities—as well as our capacities for resilience, growth, and discovery—in the most remote and delicate landscapes on our planet.

# ACKNOWLEDGMENTS

This book was a product of interactions with the landscape, the people working and living in polar regions, and those who helped me bring the story to life.

Farish A. Jenkins Jr., Bill Amaral, and Chuck Schaff, all of Harvard University, were my colleagues on my first Greenland trip in 1988. I had no idea at the time that that experience would change the course of my life, ultimately leading to this book. While each of them is now deceased, they remain very much a part of my own expeditions. I'm grateful for the confidence and patience that they had with such a neophyte. I brought Barry Lopez's *Arctic Dreams* with me on that first Greenland trip—his voice became like a fifth crew member.

Later Greenland teams included fellow students, Danish collaborators, and senior colleagues. I learned much from each of them. Steve Gatesy and I shared a memorable hike when we discovered the Dinosaur Dance Floor, and the conversations we had about our shared struggles have remained with me decades later. Eddie

Jenkins (Farish's brother) came along to Greenland one year and his stories and interactions with the landscape affected my own.

I first met Ted Daeschler when he wanted to start graduate school at the University of Pennsylvania, where I was on the faculty at the time. Little did I know that that collaboration and friendship would seed decades of work in Nunavut on the hunt for fossil fish with arms and legs. He and I had the privilege of coleading expeditions that included colleagues and crew: Brian Atagootak, Marcus Davis, Jason Downs, Kevin Middleton, Michael Shapiro, Scott Madsen, Kalliopi Monoyios, Becky Shearman, Jack Conrad, Corwin Sullivan, Stephen Gatesy, Brandon Kilbourne, Josh Miller, Andrew Gillis, Randy Dahn, Mark Webster, and Adam Maloof.

The Polar Continental Shelf Program of Natural Resources Canada provided expert logistical support for our fieldwork. We had the privilege of working with remarkably dedicated and talented people in both Ottawa and Resolute Bay over the years.

I am grateful for the communities of Grise Fiord and Resolute Bay for their support of my work as I am to the Hunters and Trappers Organization in each community. Sylvie LeBlanc and Doug Stenton at the Department of Culture, Language, Heritage and Art of Nunavut assisted with our permits as well as providing guidance with translating our results for local communities. At the Canadian Museum of Nature, Steve Cumbaa, Kieran Shepherd, Tetsuto Miyashita, and Scott Rufolo assisted with collections support.

I was fortunate to collaborate in Antarctica with Ted again, as well as with Adam Maloof, John Long, and Tim Senden. Sune Tamm, our mountaineer for our work on the margin of the Dry Valleys, provided special insights with his knowledge of polar ecosystems and law. Forrest McCarthy, our mountaineer during our austral summer on Deception Glacier, taught me about crevasses and search and rescue, and gave me insights for managing crews under challenging situations. I'm grateful to all.

The United States Antarctic Program (USAP) of the National Science Foundation is a group of people amazingly dedicated to polar research. Bija Sass, of Leidos (a contractor for the USAP), oversaw our logistic support in McMurdo and spent time talking with me about logistic support in general as I was beginning this book.

I am fortunate to have had a wonderful laboratory group at the University of Chicago during this project, including Shahid Ali, Melvin Bonilla, Matteo Fabbri, Yara Haridy, Emily Hillan, Dylan Joekel, Mirna Marinic, Katie Mika, Sam Norris, Sheri Kult Perry, Igor Schneider, Gayani Senevirathne, Neelima Sharma, Tom Stewart, and Julius Tabin. I thank all of them for their patience with me during this undertaking.

I'm deeply grateful to my deans at the University of Chicago Biological Sciences Division and University of Chicago Medicine—Mark Anderson, Iris Romero, Conrad Gilliam, and Kenneth Polonsky—for their support of me and my work.

My University of Chicago colleagues Doug MacAyeal and Kelly Brunt got me started with my work in Antarctica. Doug provided important insights and inspiration during the early stages of my research on ice.

Kevin Righter, Philipp Heck, and Ralph Harvey generously got me off to a solid start on Antarctic meteorites. Sridhar Anandakrishnan, Doug MacAyeal, and Eric Rignot graciously offered me time to interview them to change the way I see glacial ice. John Priscu offered references and advice on the microbial biota. Maureen Raymo helped me understand rock weathering and, just as important, the struggles women have faced in the field, even when they are in leadership positions. Byron Adams offered generous support for me to understand the creatures living in the Dry Valleys, even sharing pictures from his own field studies. Seth Finnegan helped me understand glaciations in the distant past, while my colleague

Zeray Alemesged tutored me and provided references on the relationship between glaciation and human evolution. Kapi Monoyios read sections of this book and gave comments with her sharp eye for narrative and detail.

Lisa Frenette and Larry Audlaluk provided insights about conveying the experiences of Indigenous Peoples in the Arctic in general, and in the community of Grise Fiord in particular. I am grateful to both of them.

My agent, Katinka Matson, at Brockman, gave critical advice at the onset of the project and helped me find a home at Dutton. Stephen Morrow got me started on this project and prodded me to think about landscapes in my writing. Emi Ikkanda helped me push the manuscript to completion with insight, close readings, and patience. I'm grateful to the entire team at Dutton for their expert work on this book, including Christine Ball, John Parsley, Grace Layer, Hannah Poole, Nicole Jarvis, Jason Booher, Lorie Pagnozzi, Nancy Resnick, Melissa Solis, Clare Shearer, Alice Dalrymple, and Gaelyn Galbreath. My editor at Oneworld in the UK, Sam Carter, provided encouragement, advice, and expert editorial comment. Thanks to all.

Family are the ones who bear the biggest cost of polar work. My children, Nathaniel and Hannah, have tolerated long absences while I'm in the field and a distracted dad while I'm writing at home. My wife, Michele, has endured decades of my disappearances to both polar regions during major life events, almost too long to enumerate individually. My expeditions and this book would not have been possible without their support and love.

# NOTES AND FURTHER READING

## PROLOGUE

The Greenland expedition in 1988 was the first of eight trips over twelve years. The initial team included Professor Farish A. Jenkins Jr., Charles (Chuck) Schaff, and William (Bill) Amaral, of the Vertebrate Paleontology Department of Harvard's Museum of Comparative Zoology; and me. Later crews had as many as eleven people. The goal was to explore the Triassic rocks of East Greenland for fossil vertebrates, and representative publications of the work are: F. A. Jenkins, N. H. Shubin et al., "Late Triassic continental vertebrates and depositional environments of the Fleming Fjord Formation, Jameson Land, East Greenland," *Meddelelser om Grønland, Geoscience* 32 (1994): 1–24; F. A. Jenkins, S. M. Gatesy, N. H. Shubin, and W. W. Amaral, "Haramiyids and Triassic mammalian evolution," *Nature* 385 (1997): 715–718; F. A. Jenkins, N. H. Shubin, S. M. Gatesy, and K. Padian, "A diminutive pterosaur (Pterosauria: eudimorphodontidae) from the Greenlandic Triassic," *Bulletin of the Museum of Comparative Zoology* 156 (2001): 151–170.

The International Polar Years began in 1882–1883 under the inspiration of Austrian scientist and naval officer Karl Weyprecht. He realized that the poles held secrets on the workings of weather, climate, and geophysics and, following one of the major themes behind polar science, believed that only a collaborative international approach could solve the big problems. A history of the International Polar Years is here: http://icestories.exploratorium.edu/dispatches/big-ideas/international-polar-year/index.html.

There have been three International Polar Years since that first one: 1932–1933, 1957–1958, and 2007–2008. The one in 1957–1958 was also called the International

Geophysical Year. That event covered not only polar science but also space research and geophysics more generally. Teams were sent across polar regions, including the South Pole, for scientific analyses. A retrospective of its achievements can be found online at the National Science Foundation (https://new.nsf.gov /science-matters/celebrating-65th-anniversary-international) and NASA (https://www.nasa.gov/centers-and-facilities/johnson/65-years-ago-the-international-geophysical-year-begins/). Archives of these meetings are in the Eisenhower Presidential Library (https://www.eisenhowerlibrary.gov/research/online-documents/international-geophysical-year-igy) and the National Academy of Sciences (https://www.nasonline.org/about-nas/history/archives/milestones-in-NAS-history/the-igy.html).

The Antarctic Treaty was signed in 1957 and went into effect in 1961 (https://www.ats.aq/e/antarctictreaty.html). The main articles are: "1. Antarctica shall be used for peaceful purposes only; 2. Freedom of scientific investigation in Antarctica and cooperation toward that end . . . shall continue; 3. Scientific observations and results from Antarctica shall be exchanged and made freely available." Mineral resources and environmental considerations appeared in a treaty signed in 1991 known as the Madrid Protocol (https://www.ats.aq/e/protocol.html).

The Harvard College student interviews appeared in the documentary *A Private Universe* (https://www.learner.org/series/a-private-universe/1-a-private-universe/). The video makes for depressing viewing for science educators.

Reasons for the extreme cold of polar regions are explained here: https://www.forbes.com/sites/marshallshepherd/2016/12/19/the-science-of-why-the-north-pole-is-cold/?sh=472ca85f57d9.

I discuss the discovery of *Tiktaalik roseae* in Neil Shubin, *Your Inner Fish: A Journey into the 3.5-Billion-Year History of the Human Body* (New York: Vintage, 2009).

## CHAPTER 1: ICE IS HOT

The United States Antarctic Program is run by the National Science Foundation. Researchers wanting to work on the ice apply, and their proposal is vetted scientifically through peer review. A later review probes the project's logistical concerns, because not all good scientific projects are feasible. Updates on research progress as well as webcams from McMurdo and South Pole Stations are here: https://www.usap.gov/. Details on McMurdo Station are found here: https://www.nsf.gov/geo/opp/support/mcmurdo.jsp.

Big Red is made by Canada Goose. You can purchase a civilian version in different colors (Big Gray?) here: https://www.canadagoose.com/us/en/expedition-parka-heritage-4660M.html.

This link shows why Antarctica is colder than the Arctic: https://www.climate.gov/news-features/features/antarctica-colder-arctic-it%E2%80%99s-still-losing-ice.

The glacial theories of Rendu, Forbes, Kelvin, and Tyndall are discussed in: J. Hollier and A. Hollier, "The glacier theory of Louis Rendu (1789–1859) and the Forbes–Tyndall controversy," *Earth Sciences History* 35 (2016): 346–353; and

D. Gross, "Plasticity, Viscosity, Regelation—Remarks on the bitter Forbes–Tyndall Controversy," *Proceedings in Applied Mathematics and Mechanics* 20 (2020): 1–2 (open access at: https://onlinelibrary.wiley.com/doi/pdf/10.1002/pamm.202000158).

Margaret Bradshaw's pioneering studies, as well as her work in the area of Deception Glacier, are covered in a book by one of my Antarctic teammates, Professor John Long. His book paints a vivid picture of camp life on the ice in Antarctica as well as the teams who first investigated the region: John Long, *Mountains of Madness: A Scientist's Odyssey in Antarctica* (Washington, D.C.: Joseph Henry Press, 2001).

For the basics on glaciers, their dangers, and where to see them safely, see: https://www.backpacker.com/stories/nature/phenomenon-crevasses/. The obituary of Gordon Hamilton, the scientist who tragically perished in the Antarctic crevasse, is here: https://www.nytimes.com/2016/10/25/science/gordon-hamilton-climate-scientist-dies-accident-antarctica.html. Professor Tim Senden of the Australian National University was my colleague who descended into our small training crevasse.

The National Snow and Ice Center has excellent background information about how glacial ice forms as it gets compressed every year: https://nsidc.org/learn/parts-cryosphere/glaciers. Antarcticglaciers.org is one-stop shopping for information about all things glacial, including the formation of glacial ice: https://www.antarcticglaciers.org/glacier-processes/from-snow-to-glacier-ice/. The site has a large and ever-growing account of glacial research and up-to-date references.

The wind-sculpted rocks are known as ventifacts. A recent study on the ways in which high glacial winds sculpt them is here: J. Ross Mackay and C. R. Burn, "A long-term field study (1951–2003) of ventifacts formed by katabatic winds at Paulatuk, western Arctic coast, Canada," *Canadian Journal of Earth Sciences* 42 (2005): 1615–1635.

For a popular site on the Dry Valleys of Antarctica, including excellent photos, see: https://www.coolantarctica.com/Antarctica%20fact%20file/antarctica%20environment/dry-valleys-blood-falls-don-juan-pond.php. For a history of the Dry Valleys, see: https://communities.springernature.com/posts/when-were-the-mcmurdo-dry-valleys-of-antarctica-last-wet. A classic paper on the history of the Dry Valleys is: D. Sugden, G. Denton, and D. Marchant, "Landscape evolution of the Dry Valleys, Transantarctic Mountains: tectonic implications," *Journal of Geophysical Research* 100 (1995): 9949–9967.

Eric Rignot has published multiple papers on the movement of the ice sheets using satellite data. The ones I use in this chapter are: E. Rignot, J. Mouginot, and B. Scheuchl, "Ice flow of the Antarctic ice sheet," *Science* 333 (2011): 1427–1430; E. Rignot, "Changes in ice dynamics and mass balance of the Antarctic ice sheet," *Philosophical Transactions of the Royal Society A* 364 (2011): 1637–1655; and E. Rignot, J. Bamber, M. van den Broeke, et al., "Recent Antarctic ice mass loss from radar interferometry and regional climate modelling," *Nature Geoscience* 1 (2008): 106–110, https://doi.org/10.1038/ngeo102/.

For a primer on what sets the mass of a glacier see: https://www.antarcticglaciers.org/glacier-processes/mass-balance/.

The Jet Propulsion Laboratory (JPL) maintains a site on the background and progress of the GRACE missions (https://grace.jpl.nasa.gov/mission/grace/). A video primer on what the missions say about changes to the ice is here: https://gracefo.jpl.nasa.gov/resources/16/grace-mission-measures-global-ice-mass-changes/. A recent paper by Eric Rignot and his team on these results is: I. Velicogna et al., "Continuity of ice sheet mass loss in Greenland and Antarctica from the GRACE and GRACE Follow-On missions," *Geophysical Research Letters* 47 (2020): 1–8.

Two fabulous books on the ice ages are: David Archer, *The Long Thaw: How Humans Are Changing the Next 100,000 Years of Earth's Climate* (Princeton, NJ: Princeton University Press, 2016); and John Imbrie and Katherine Palmer Imbrie, *Ice Ages: Solving the Mystery* (Cambridge, MA: Harvard University Press, 1986).

A recent paper on ice and sea-level changes over the past 66 million years is: K. G. Miller et al., "Cenozoic sea-level and cryospheric evolution from deep-sea geochemical and continental margin records," *Science Advances* 15 (2020): 1–15.

## CHAPTER 2: HIDDEN WORLDS

The myths that surround Hitler's retreat to an Antarctic base are covered in: C. Summerhayes and P. Beeching, "Hitler's Antarctic base: the myth and the reality," *Polar Record* 43 (2007): 1–21. Ladislas Szabó's books that catalyzed the rumor of a Nazi base are: Ladislas Szabó, *Je sais sue Hitler est Vivant* (Buenos Aires: SFELT, 1947), and Ladislas Szabó, *Antarctic Exploration: Hitler's Escape to South America and Secret Bases in Antarctica*, Historic Series (Cheltenham, UK: Reardon Publishing, Cheltenham, 2022 [reprint]). A popular description of Szabó's myth is at: https://www.coolantarctica.com/Community/antarctic-mysteries-hitlers-secret-base.php.

H. P. Lovecraft envisioned a fictional station that, by coincidence, was only miles from where my team worked on Deception Glacier in Antarctica in 2019. His book has been reprinted many times. See, for example, H. P. Lovecraft, *At the Mountains of Madness* (Floyd, VA: Positronic Press, 2022 [reprint]).

Peter Kropotkin's autobiography has been translated into English: Peter Kropotkin, *Memoirs of a Revolutionist* (New York: Dover Press, 2011). His geographical theories are laid out in his seven-hundred-page treatise: Peter Kropotkin, *Investigation of Quaternary Period, Notes of Russian Geographical Society* 7 (1876), 717 pp. [in Russian]. As I am unable to read Russian, and there is not a translation available, I have not read this text and have referred to it using secondary sources such as https://embryo.asu.edu/pages/petr-alekseevich-kropotkin-1842-1921/.

*Wired* published a readable account of Vostok Station in 2004: https://www.wired.com/2000/04/vostok/. The station is now getting an upgrade: https://polarjournal.ch/en/2022/01/24/construction-start-of-new-russian-vostok-station-in-antarctica/.

## NOTES AND FURTHER READING

The legend about the Antarctic Soviet banning chess after an axe attack appears in: https://canadiangeographic.ca/articles/how-antarctic-isolation-affects-the-mind/ as well as Emma Barrett and Paul Martin, *Extreme: Why Some People Thrive at the Limits* (Oxford: Oxford University Press, 2016). I have been unable to find original Soviet sources documenting the event; hence, I've labeled the story as legend.

Andrei Kapitza's life was recounted in his obituaries from 2011, see: https://www.telegraph.co.uk/news/obituaries/science-obituaries/8725763/Andrei-Kapitsa.html and https://ges.rgo.ru/jour/article/viewFile/118/119/.

Kapitsa's original papers on Subglacial Lake Vostok are: A. P. Kapitsa, "Dynamics and morphology of ice sheet in the central part of East Antarctica," *Transactions of Soviet Antarctic Expedition* 19 (1961): 93 pp. [in Russian]; A. P. Kapitsa et al., "A large deep freshwater lake beneath the ice of central East Antarctica," *Nature* 381 (1996): 684–686.

Accounts of Subglacial Lake Vostok in the popular press include: https://www.livescience.com/38652-what-is-lake-vostok.html; https://www.americanscientist.org/article/antarcticas-lake-vostok; https://science.howstuffworks.com/environmental/earth/geophysics/lake-vostok.htm; and https://www.wired.com/2012/02/lake-vostok-drilled/.

Primary sources on the geology and structure of Subglacial Lake Vostok include: G. L. Leitchenkov, A. V. Antonov, P. I. Luneov, and V. Y. Lipenkov, "Geology and environments of subglacial Lake Vostok," *Philosophical Transactions of the Royal Society A: Mathematical, Physical and Engineering Sciences* 374 (2016), 20140302; and M. Studinger et al., "Ice cover, landscape setting, and geological framework of Lake Vostok, East Antarctica," *Earth and Planetary Science Letters* 205 (2003): 195–210.

Research on life in Subglacial Lake Vostok is reported in: J. C. Priscu et al., "Geomicrobiology of subglacial ice above Lake Vostok, Antarctica," *Science* 286 (1999): 2141–2144; M. J. Siegert et al., "Physical, chemical and biological processes in Lake Vostok and other Antarctic subglacial lakes," *Nature* 414 (2001): 603–609; B. C. Christner et al., "Limnological conditions in subglacial lake Vostok, Antarctica," *Limnology and Oceanography* 51 (2006): 2485–2501; and M. J. Siegert et al., "Antarctic subglacial lake exploration: first results and future plans," *Philosophical Transactions of the Royal Society A: Mathematical, Physical and Engineering Sciences* 374 (2016): 20140466.

Background on John Priscu and the search for life in subglacial Antarctic lakes is here: https://www.quantamagazine.org/john-priscu-finds-life-in-antarcticas-frozen-lakes-20200720/.

Work at Lake Whillans is described in: J. C. Priscu, R. D. Powell, and S. Tulaczyk, "Probing subglacial environments under the Whillans Ice Stream," *Eos, Transactions American Geophysical Union* 91 (2010): 253–254; J. C. Priscu et al., "A microbiologically clean strategy for access to the Whillans Ice Stream subglacial environment," *Antarctic Science* 25 (2013): 637–647; and D. Fox, "Antarctica's Secret Garden," *Nature* 512 (2014): 244.

Fieldwork at Subglacial Lake Mercer began in earnest in 2018 (we shared lab space with Priscu's large team, with the acronym SALSA, for Subglacial Antarctic Lakes Scientific Access). Initial results are in these three publications: J. C. Priscu et al., "Scientific access into Mercer Subglacial Lake: scientific objectives, drilling operations and initial observations," *Annals of Glaciology* 62 (2021): 340–352; D. Fox, "Life below the ice," *Nature* 564 (2018): 180–182; C. L. Davis et al., "Biogeochemical and historical drivers of microbial community composition and structure in sediments from Mercer Subglacial Lake, West Antarctica," *ISME Communications* 3 (2023): article 8.

Professor Scott Rodgers's scientific paper describing surprisingly diverse life in Subglacial Lake Vostok is: https://journals.plos.org/plosone/article?id=10.1371/journal.pone.0067221 (open access).

*The Antarctic Sun* is McMurdo Station's newspaper, often offering features on some of the working groups there. This article is on the remarkable people of the "Helo Ops" group, including a podcast and interviews with personnel: https://antarcticsun.usap.gov/features/4336/.

The general geology of Blood Falls is discussed in: https://geologyscience.com/gallery/geological-wonders/blood-falls-antarctica/. Scientific analyses of Blood Falls are published in: https://www.frontiersin.org/articles/10.3389/fspas.2022.843174/full and J. Badgley et al., "An englacial hydrologic system of brine within a cold glacier: Blood Falls, McMurdo Dry Valleys, Antarctica," *Journal of Glaciology* 63 (2017): 387–400. The microbial biota is discussed in: J. A. Mikucki and J. C. Priscu, "Bacterial diversity associated with Blood Falls, a subglacial outflow from the Taylor Glacier, Antarctica," *Applied and Environmental Microbiology* 73 (2007): 4029–4039. A good popular account is in: https://www.popsci.com/antarctica-blood-falls/.

The viral spillover risk in Lake Hazen, Canada, is discussed in: A. Lemieux, G. A. Colby, A. J. Poulain, and S. Aris-Brosou, "Viral spillover risk increases with climate change in High Arctic lake sediments," *Proceedings of the Royal Society B* 289 (2022): 20221073. The general problem of glacial melt and viral spillover is recounted here: R. Varghese, P. Patel, D. Kumar, and R. Sharma, "Climate change and glacier melting: risks for unusual outbreaks?," *Journal of Travel Medicine* 30 (2023): taad015.

This paper describes the 15,000-year-old viruses from the Guliya Ice Cap: Z. P. Zhong et al., "Glacier ice archives nearly 15,000-year-old microbes and phages," *Microbiome* 9 (2021): 1–23.

Analyses of viruses in ancient Siberian permafrost are described in: J. M. Alempic et al., "An update on eukaryotic viruses revived from ancient permafrost," *Viruses* 15 (2023): 564.

The 2016 anthrax outbreak in Siberia, and its relation to climate change, is discussed here: E. Ezhova et al., "Climatic factors influencing the anthrax outbreak of 2016 in Siberia, Russia," *EcoHealth* 18 (2021): 217–228. Joshua Yaffa, writing in *The New Yorker*, discusses melting permafrost in: https://www.newyorker.com/magazine/2022/01/17/the-great-siberian-thaw/.

The molecular mechanism of bacterial reanimation after being in suspended animation is described here: Y. Gao et al., "Bacterial spore germination receptors are nutrient-gated ion channels," *Science* 380, no. 3 (2023): 387–391. A popular account of the story is in: https://hms.harvard.edu/news/how-dormant-bacteria-return-life.

The experiment showing how rock erosion can create nutrient conditions necessary to support life is: B. Gill-Olivas et al., "Subglacial erosion has the potential to sustain microbial processes in Subglacial Lake Whillans, Antarctica," *Communications Earth and Environment* 2 (2021): 134.

## CHAPTER 3: COLD COMFORTS

Arctic willow (*Salix arctica*) is a subject of a sumptuous essay by Barry Lopez in *Arctic Dreams* (New York: Scribner, 1986). Recent studies of its ecology and variation include: N. Boulanger-Lapointe et al., "Population structure and dynamics of Arctic willow (*Salix arctica*) in the High Arctic," *Journal of Biogeography* 41 (2014): 1967–1978; A. M. Pajunen, "Environmental and biotic determinants of growth and height of Arctic willow shrubs along a latitudinal gradient," *Arctic, Antarctic, and Alpine Research* 44 (2009): 478–485; and K. S. Christie, R. W. Ruess, M. S. Lindberg, and C. P. Mulder, "Herbivores influence the growth, reproduction, and morphology of a widespread Arctic willow," *PLoS One* 9 (2014): e101716.

A general source for information on the Arctic woolly bear caterpillar (*Gynaephora groenlandica*), including its thermoregulatory biology, is at: https://inaturalist.ca/taxa/452615-Gynaephora-groenlandica. The ecology and physiology of the species are discussed in: V. A. Bennett, "Ecological physiology of the Arctic woollybear caterpillar *Gynaephora groenlandica* (Lepidoptera: Lymantriidae)," PhD diss., Miami University (available at: https://www.proquest.com/docview/304699725?origsite=gscholar&fromopenview=true&sourcetype=Dissertations%20&%20Theses).

The folklore about the ways in which caterpillars can predict weather is discussed by the National Weather Service (https://www.weather.gov/arx/woollybear) and the *Farmers' Almanac* (https://www.farmersalmanac.com/caterpillars-and-a-rough-winter).

Anna Bågenholm's story has been documented in multiple media, including articles like https://info.soshydration.com.au/anna-bagenholm-froze-to-death-but-lived/ and https://explorersweb.com/great-survival-stories-anna-bagenholm/ as well as video (see the BBC's at https://www.youtube.com/watch?v=jLr15BBBtrc).

Arthur DeVries's scientific journey is covered in https://www.scientia.global/professor-arthur-devries-cold-ice-antifreeze-proteins-polar-fishes/ and https://www.nytimes.com/1997/04/15/science/fighting-cold-at-both-poles.html. His seminal papers are: A. L. DeVries and D. E. Wohlschlag, "Freezing resistance in some Antarctic fishes," *Science* 163 (1969): 1073–1075; A. L. DeVries, "Glycoproteins as biological antifreeze agents in antarctic fishes," *Science* 1172 (1971): 152–155;

and J. A. Raymond and A. L. DeVries, "Adsorption inhibition as a mechanism of freezing resistance in polar fishes," *Proceedings of the National Academy of Sciences* 74 (1977): 2589–2593. The molecular workings of the antifreeze proteins and their use in VanEELA ice cream are described in: https://www.openaccessgovernment.org/role-antifreeze-proteins-freezing-avoidance-polar-fishes/40377/. General adaptations that animals have evolved to survive the cold are in: K. B. Storey and J. M. Storey, "Frozen and alive," *Scientific American* 263, no. 6 (1990): 92–97.

The role of the sugar trehalose in antifreeze functions in insects is discussed in: X. Wen et al., "Antifreeze proteins govern the precipitation of trehalose in a freezing-avoiding insect at low temperature," *Proceedings of the National Academy of Sciences* 113 (2016): 6683–6688; and I. Kratochvílová et al., "Theoretical and experimental study of the antifreeze protein AFP752, trehalose and dimethyl sulfoxide cryoprotection mechanism: correlation with cryopreserved cell viability," *RSC Advances* 7 (2017): 352–360.

General mechanisms of survival for small worms in Antarctic conditions can be found in: D. A. Wharton, G. Goodall, and C. J. Marshall, "Freezing survival and cryoprotective dehydration as cold tolerance mechanisms in the Antarctic nematode *Panagrolaimus davidi*," *Journal of Experimental Biology* 206 (2003): 215–221. A broad treatment of adaptation of small invertebrates to extremes of cold and dry is: D. A. Wharton and C. J. Marshall, "How do terrestrial Antarctic organisms survive in their harsh environment?," *Journal of Biology* 8 (2009): 1–4.

A review of polar adaptations in vertebrates is in Arnoldus Schytte Blix, *Arctic Animals and Their Adaptations to Life on the Edge* (Oslo: Tapir Academic Press, 2005); and A. S. Blix, "Adaptations to polar life in mammals and birds," *Journal of Experimental Biology* 219 (2016): 1093–1105.

A general account of the natural history of musk ox (*Ovibos mochatus*) is here: https://www.coolantarctica.com/Antarctica%20fact%20file/wildlife/Arctic_animals/musk-ox.php. To learn more about the remarkable natural fabric qiviut, go to https://www.muskoxfarm.org/qiviut. There you can buy qiviut products and even adopt a musk ox.

A general source on polar bear biology, written by one of the scientific leaders in the field, is: Andrew Edward Derocher, *Polar Bears: A Complete Guide to Their Biology and Behavior* (Baltimore: Johns Hopkins University Press, 2012).

The parallel deployment of arteries and veins, and the ways polar bears conserve heat and other important biological entities, is called a countercurrent exchange system. See K. Schmidt-Nielsen, "Countercurrent systems in animals," *Scientific American* 244 (1981): 18–129, for details on their physiology and variation.

There are two classic "rules" in biology (more like general regularities than biological laws): Bergmann's rule and Allen's rule. The former is named after Carl Bergmann, who in 1847 argued that in a widespread group of animals, larger populations will be found in cold habitats and smaller ones in warmer locales. C. Bergmann, "Über die Verhältnisse der Wärmeökonomie der Thiere zu ihrer grösse," *Gottinger Studien* 3 (1847): 595–708. Allen's rule, named for Joel Asaph Allen in 1877, states that creatures that live in colder environments will have smaller appendages (http://people.wku.edu/charles.smith/biogeog/ALLE1877.htm). There have been countless studies of these effects in different species. See reviews in: S. Meiri and T. Dayan, "On the validity of Bergmann's rule," *Journal of Biogeography* 30 (2003): 331–351; and R. L. Nudds and S. A. Oswald, "An interspecific test of Allen's rule: evolutionary implications for endothermic species," *Evolution* 61 (2007): 2 [2839] 839–2848.

The ways in which these "rules" work in humans has been controversial. See, for example: F. Foster and M. Collard, "A reassessment of Bergmann's rule in modern humans," *PloS One* 8 (2013): 72269; and L. Betti, S. J. Lycett, N. von Cramon-Taubadel, and O. M. Pearson, "Are human hands and feet affected by climate? A test of Allen's rule," *American Journal of Physical Anthropology* 158 (2015): 132–140.

A readable, and often funny, discussion of the importance of biological clocks is in John D. Palmer, *The Living Clock: The Orchestrator of Biological Rhythms* (New York: Oxford University Press, 2002). The three labs that cloned clock genes and explored their biological workings were Jeffrey Hall (Brandeis University); Michael Rosbach (Brandeis University); and Michael Young (Rockefeller University). There is a large body of work on clocks in diverse species. See: J. Dunlap, "Molecular basis for circadian clocks," *Cell* 96 (1999): 271–290; M. Rosbach, "Implications of multiple circadian clock origins," *PLoS Biology* 7 (2009): 17–25; and S. Panda, J. B. Hogenesch, and S. A. Kay, "Circadian rhythms from flies to human," *Nature* 417 (2022): 329–335.

The effect of twenty-four hours of daylight or darkness on animal biology is reviewed in: G. Bloch, B. M. Barnes, M. P. Gerkema, and B. Helm, "Animal activity around the clock with no overt circadian rhythms: patterns, mechanisms and adaptive value," *Proceedings of the Royal Society B: Biological Sciences* 280 (2013): 20130019.

For a review of the role of aerobic capacity and thermal adaptation in human evolution, see: D. E. Lieberman, "Human locomotion and heat loss: an evolutionary perspective," *Comprehensive Physiology* 5 (2011): 99–117.

Greenlandic Inuit have adaptations to local diet and environment as discussed in: M. Fumagalli et al., "Greenlandic Inuit show genetic signatures of diet and climate adaptation," *Science* 349 (2015): 1343–1347. The National Institutes of Health summarized this paper for a general audience: https://www.nih.gov/news-events/nih-research-matters/genetic-adaptations-diet-climate. More recent studies have found other adaptations: N. Senftleber et al., "Genetic study of the Arctic CPT1A variant suggests that its effect on fatty acid levels is modulated by traditional Inuit diet," *European Journal of Human Genetics* 28 (2020): 1592–1601.

If you want to lose weight, get cold. Shivers burn calories, and brown fat, which also burns calories, gets deposited. Brown-fat metabolism, and its relationship to cold environments, is discussed in: B. Cannon and N. Jan, "Brown adipose tissue: function and physiological significance," *Physiological Reviews* 84

(2004): 77–359; and, in an account for a more general audience, see https://www.scientificamerican.com/article/supercharging-brown-fat-to-battle-obesity/.

The study that chilled sleeping volunteers was called the ICEMAN study and is described here: https://www.garvan.org.au/news-resources/news/the-iceman-study-how-keeping-cool-could-spur-metabolic-benefits, with the results reported in: P. Lee et al., "Temperature-acclimated brown adipose tissue modulates insulin sensitivity in humans," *Diabetes* 63 (2014): 3686–3698.

Other studies have looked at brown-fat metabolism and synthesis in cold environments, including in male swimmers: S. Søberg et al., "Altered brown fat thermoregulation and enhanced cold-induced thermogenesis in young, healthy, winter-swimming men," *Cell Reports Medicine* 2, no. 10 (2021); as well as other populations: K. Y. Chen, "Brown fat activation mediates cold-induced thermogenesis in adult humans in response to a mild decrease in ambient temperature," *Journal of Clinical Endocrinology and Metabolism* 98, no. 7 (2013): E1218–E1223.

The effects of "free running" biological clocks in humans working in Antarctica is in: D. J. Kennaway and C. F. Van Dorp, "Free-running rhythms of melatonin, cortisol, electrolytes, and sleep in humans in Antarctica," *American Journal of Physiology: Regulatory, Integrative and Comparative Physiology* 260 (1991): R1137–R1144.

The psychological challenges of working in Antarctica have been documented in: Julian Sancton, *Madhouse at the End of the Earth: The Belgica's Journey into the Dark Antarctic Night* (New York: Crown, 2021); L. Palinkas, "Going to extremes: the cultural context of stress, illness, and coping in Antarctica," *Social Science and Medicine* 35 (1992): 651–664; and Donna Oliver, "Some psychological effects of isolation and confinement in an Antarctic winter-over group," PhD diss., United States International University (available at: https://www.proquest.com/docview/302983797?parentSessionId=8AbpyccmVmV4AN4cHkV0aTNxSsSr2WWqQ4vHspSkRG0%3D&sourcetype=Dissertations%20&%20Theses.

"Adventure psychology" is a subfield that looks at the mental conditions that arise from extreme fieldwork, as well as strategies to cope with the challenges. See: Paula Reid and Eric Brymer, *Adventure Psychology: Going Knowingly into the Unknown* (London: Routledge, 2022); N. Smith, F. Kinnafick, and B. Saunders, "Coping strategies used during an extreme Antarctic expedition," *Journal of Human Performance in Extreme Environments* 13 (2017): 1–9; P. Reid and H. Kampman, "Exploring the psychology of extended-period expeditionary adventurers: going knowingly into the unknown," *Psychology of Sport and Exercise* 46 (2020): 101608.

Jack Stuster's book *Bold Endeavors* (Annapolis: Naval Institute Press, 1996) explores the lessons in psychology, group dynamics, and leadership that stem from polar expeditions and space exploration.

## CHAPTER 4: OUT OF THIS WORLD

Sir Douglas Mawson wrote about his 1911–1914 expedition in *The Home of the Blizzard: An Australian Hero's Tale of Antarctic Discovery and Adventure*

(Cambridge, MA: Wakefield Press, 2010 [reprint]). A harrowing third-party account is in *Smithsonian*: https://www.smithsonianmag.com/history/the-most-terrible-polar-exploration-ever-douglas-mawsons-antarctic-journey-82192685/.

Frank Bickerton's biography is: Stephen Haddelsey, *Born Adventurer: The Life of Frank Bickerton, Antarctic Pioneer* (Cheltenham, UK: The History Press, 2005). The meteorite he discovered is known as the Adelie Land Meteorite and is in the collection of the Australian Museum at: https://australian.museum/learn/minerals/mineral-factsheets/adelie-land-meteorite/. Bickerton's field notes are in the Scott Polar Research Center, with the quote on the meteorite discovery at: SPRI MS 1509, "A log of the Western Journey, January 11, 1913" (https://www.spri.cam.ac.uk/library/catalogue/sprilib/antarctica/).

The interchange between Professor Gorai and Dr. Masaru Yoshida appears in Yoshida's highly informative and entertaining account: M. Yoshida, "Discovery of the Yamato meteorites in 1969," *Polar Science* 3 (2010): 272–284.

William Cassidy's book *Meteorites, Ice, and Antarctica: A Personal Account* (Cambridge: Cambridge University Press, 2012) discusses the birth of the Antarctic Search for Meteorites Program (ANSMET) and the ways we study the meteorites themselves.

The website for ANSMET is: https://caslabs.case.edu/ansmet/. You can follow recent expeditions and see photos. Want to join them? It's a highly competitive application process, but information on applying to be an Antarctic meteorite hunter is here: https://caslabs.case.edu/ansmet/faqs/. Information on the meteorite collection itself is at: https://curator.jsc.nasa.gov/antmet/.

Information on the Crary Lab at McMurdo Station is at: https://www.nsf.gov/geo/opp/support/crarylab.jsp.

The dynamics of ice, and the ways in which it collects meteorites, is discussed in Cassidy, *Meteorites, Ice, and Antarctica.* He also discusses aluminum isotopes and the ways they inform us about how long meteorites have sat in the ice.

NASA's Jet Propulsion Laboratory holds information about the Nakhla Meteorite: https://www2.jpl.nasa.gov/snc/nakhla.html. *Smithsonian* magazine tells the larger story of the meteorite and its fall: https://www.smithsonianmag.com/smithsonian-institution/one-hundred-years-ago-today-a-mars-meteorite-fell-in-a-blaze-23722099/.

Meteorite EETA79001 is discussed in Cassidy, *Meteorites, Ice, and Antarctica.*

Meteorite ALH84001, and the Martian life story, is discussed in BBC Science Focus: https://www.sciencefocus.com/space/life-on-mars-the-story-of-meteorite-alh84001. The original claim for life on Mars is in: D. S. McKay et al., "Search for past life on Mars: possible relic biogenic activity in Martian meteorite ALH84001," *Science* 273 (1996): 924–930. Another claim for Martian life is in: K. L. Thomas-Keprta et al., "Magnetofossils from ancient Mars: a robust biosignature in the Martian meteorite ALH84001," *Applied and Environmental Microbiology* 68 (2002): 3663–3672. An evaluation of the evidence is in: E. K. Gibson Jr. et al., "Life on Mars: evaluation of the evidence within Martian meteorites

ALH84001, Nakhla, and Shergotty," *Precambrian Research* 106, nos. 1–2 (2002): 15–34. A retrospective of the debate is at: https://www.space.com/33690-allen-hills-mars-meteorite-alien-life-20-years.html/.

The story of cosmic dust accumulating on house rooftops is in: M. J. Genge, J. Larsen, M. Van Ginneken, and M. D. Suttle, "An urban collection of modern-day large micrometeorites: evidence for variations in the extraterrestrial dust flux through the Quaternary," *Geology* 45 (2017): 119–122. It is discussed in the popular press in https://www.nytimes.com/2017/03/10/science/space-dust-on-earth.html and https://www.cbc.ca/radio/quirks/september-30-2018-1.4312266/your-rooftop-could-be-sprinkled-with-cosmic-star-dust-1.4312284/.

Interstellar dust in Antarctic ice is announced and analyzed here: D. Koll et al., "Interstellar Fe 60 in Antarctica," *Physical Review Letters* 123 (2002): 072701, with a summary for a general audience at https://www.space.com/interstellar-dust-in-antarctic-snow.html/.

For a synthesis of what meteorites tell us about the origin of the solar system, see: Caroline Smith, Sarah Russell, and Natash Almeida, *Meteorites: The Story of Our Solar System* (Ontario: Firefly Books, 2019).

## CHAPTER 5: A BRIEF HISTORY OF ICE

Nansen's original account of his voyage in the *Fram* is: Fridjof Nansen, *Farthest North: Being the Record of a Voyage of Exploration of the Ship* Fram *1893–96, and of a Fifteen Months' Sleigh Journey by Dr. Nansen and Lieut. Johansen* (New York: Harper & Brothers Publishers, 1898). The Nobel Prize Foundation has an excellent short biography of Nansen: https://www.nobelprize.org/prizes/peace/1922/nansen/biographical/.

The second *Fram* expedition wintered on southern Ellesmere Island, making scientific observations, including the discovery of Devonian fossil fish: Otto Sverdrup, "The second Norwegian polar expedition in the *Fram*, 1898–1902," *Scottish Geographical Magazine* 19, no. 7 (1903): 337–353; and Olaf Holtedahl, *Summary of Geological Results: With Geological Map, 6 Plates and 4 Figures in the Text,* Vol. 36 (Oslo: AW Brøgger, 1917).

A biography of Lady Jane Franklin is: Alison Alexander, *The Ambitions of Lady Jane Franklin: Victorian Lady Adventurer* (London: Allen and Unwin, 2016).

The theory of Open Polar Seas has a long history, see: J. K. Wright, "The Open Polar Sea," *Geographical Review* 43 (1953): 338–365.

Elisha Kent Kane's monographs on his two Grinnell Expeditions make for lively reading for both his prose and the evocative woodcuts depicting life aboard ship and on the ice: Elisha Kent Kane, *The US Grinnell Expedition in Search of Sir John Franklin: A Personal Narrative* (Philadelphia: Childs & Peterson, 1856); and Elisha Kent Kane, *Arctic Explorations: The Second Grinnell Expedition in Search of Sir John Franklin, 1853,' '54, '55* (Philadelphia: Childs & Peterson, 1861). Morton's quote about encountering the Open Polar Sea is in: Wright, "The Open Polar Sea."

# NOTES AND FURTHER READING

Isaac Hayes's account of his voyage is: Dr. I. I. Hayes, *The Open Polar Sea: A Narrative of a Voyage of Discovery Towards the North Pole, in the Schooner "United States"* (London: Sampson Low, Son and Marston, 1867).

Robert Falcon Scott's diaries are in: https://www.spri.cam.ac.uk/museum/diaries/scottslastexpedition/; and Robert Falcon Scott, *Scott's Last Expedition*, Classics of World Literature (London: Wordsworth Editions, 2012).

Articles detailing the extraordinary science that Scott and his team did are in: E. J. Larson, "Greater glory," *Scientific American* 304 (2011): 78–83; and https://www.theguardian.com/uk/2011/sep/24/scott-antarctic-lies-race-pole/.

Edward Wilson's notebooks, including his sketches, are in: David Wilson and Christopher Wilson, *Edward Wilson's Antarctic Notebooks* (Cheltenham, UK: Reardon, 2011). I stared at these drawings, and Wilson's notes, for hours while thinking about this book.

The description of Robert Falcon Scott's fossil plants is in: A. C. Seward, "Antarctic Fossil Plants," in *British Antarctic ("Terra Nova") Expedition, 1910*, Natural History Report, *Geology* 1, no. 1 (London: British Museum [Natural History]), 1–49. A discussion of the fossil plants that Wilson discovered on the Beardmore Glacier is in: W. G. Chaloner and P. Kenrick, "Did Captain Scott's Terra Nova Expedition discover fossil Nothofagus in Antarctica?," *The Linnean Society of London* (October 2015): 11–17. The story of *Glossopteris* and its importance is in: N. R. Cúneo, J. Isbell, E. L. Taylor, and T. N. Taylor, "The Glossopteris flora from Antarctica: taphonomy and paleoecology," *Comptes Rendus* XII ICC-P, 2 (1993): 13–40.

The recognition of glacial features in sediment layers is discussed in: W. B. Harland, K. N. Herod, and D. H. Krinsley, "The definition and identification of tills and tillites," *Earth-Science Reviews* 2 (1966): 225–256; and A. Dreimanis and C. Schlüchter, "Field criteria for the recognition of till or tillite," *Palaeogeography, Palaeoclimatology, Palaeoecology* 51 (1985): 7–14.

The ways latitude and longitude are recognized in rocks is the study of paleomagnetism: Lisa Tauxe, *Essentials of Paleomagnetism* (Berkeley: University of California Press, 2010).

The maps of the ancient Earth are depicted in: C. R. Scotese, "An atlas of Phanerozoic paleogeographic maps: the seas come in and the seas go out," *Annual Review of Earth and Planetary Sciences* 49 (2021): 679–728.

Arthur Philemon Coleman's description of the first ice age is: A. P. Coleman, "The lower Huronian ice age," *Journal of Geology* 16 (1908): 149–158.

The Snowball Earth theory is discussed in: P. F. Hoffman, A. J. Kaufman, G. P. Halverson, and D. P. Schrag, "A Neoproterozoic snowball earth," *Science* 281, no. 5381 (1998): 1342–1346; Joseph Kirshvink's original article (https://web.gps.caltech.edu/~jkirschvink/pdfs/firstsnowball.pdf); and Gabrielle Walker, *Snowball Earth: The Story of a Maverick Scientist and His Theory of the Global Catastrophe That Spawned Life as We Know It* (New York: Crown, 2007). Some of the initial disagreements about the theory are discussed in: N. Lubick, "Palaeoclimatology: snowball fights," *Nature* 417 (2002): 2–14.

Discussion of the timing of the formation of our ice-capped poles is in: R. M. DeConto and D. Pollard, "Rapid Cenozoic glaciation of Antarctica induced by declining atmospheric $CO_2$," *Nature* 421, no. 6920 (2003): 245–249; R. M. DeConto et al., "Thresholds for Cenozoic bipolar glaciation," *Nature* 455, no. 7213 (2008): 652–656; and https://www.geolsoc.org.uk/Education-and-Careers/Ask-a-Geologist/Earths-Climate/How-Long-has-Earth-had-Polar-Ice-Caps.

## CHAPTER 6: HOW TO FREEZE A POLE

Steve Gatesy and I were walking the cliffs of McKnight Berg in East Greenland. Steve's paper on trackways of the Dinosaur Dance Floor is: S. M. Gatesy, K. M. Middleton, F. A. Jenkins Jr., and N. H. Shubin, "Three-dimensional preservation of foot movements in Triassic theropod dinosaurs," *Nature* 399 (1999): 141–144.

Ébelmen's contribution to geoscience has been discussed in: R. A. Berner, "Jacques-Joseph Ébelmen, the founder of earth system science," *Comptes Rendus Geoscience* 344 (1999): 544–548; and R. A. Berner and K. Maasch, "Chemical weathering and controls on atmospheric $O_2$ and $CO_2$: fundamental principles were enunciated by J. J. Ebelmen in 1845," *Geochimica et Cosmochimica Acta* 60 (1996): 1633–1637. His original paper is: J. J. Ébelmen, "Sur les produits de la décomposition des espèces minérales de la famille des silicates," *Annales des Mines* 7 (1845): 66.

Carl Sagan's almost seventeen-minute testimony to the U.S. House of Representatives remains the single best resource on the greenhouse effect that I have seen. This video is well worth the time to watch: https://www.youtube.com/watch?v=Wp-WiNXH6hI/.

Harold Urey's biography is in his National Academy of Sciences memoir: https://www.nasonline.org/publications/biographical-memoirs/memoir-pdfs/urey-harold.pdf; and at https://chemistry.berkeley.edu/news/bomb-moon-harold-urey-alum-and-nobel-laureate-principles/.

The relationship between chemical weathering, volcanic activity, and global carbon is in: R. A. Berner, "A new look at the long-term carbon cycle," *GSA Today* 9 (1999): 1–6; R. A. Berner, "Weathering, plants, and the long-term carbon cycle," *Geochimica et Cosmochimica Acta* 56 (1992): 3225–3231; and R. A. Berner, A. C. Lasaga, and R. M. Garrels, "Carbonate-silicate geochemical cycle and its effect on atmospheric carbon dioxide over the past 100 million years," *American Journal of Science* 283 (1983): 641–683.

Maureen Raymo's original papers are: M. E. Raymo and W. F. Ruddiman, "Tectonic forcing of late Cenozoic climate," *Nature* 359 (1992): 117–122; M. E. Raymo, W. F. Ruddiman, and P. N. Froelich, "Influence of late Cenozoic mountain building on ocean geochemical cycles," *Geology* 16 (1988): 649–653; and M. E. Raymo, "The Himalayas, organic carbon burial, and climate in the Miocene," *Paleoceanography* 9 (1994): 399–404. A paper looking at the relationship between climactic cooling and uplift of the Tibetan Plateau is: C. Garzione, "Surface uplift of Tibet and Cenozoic global cooling," *Geology* 36 (2008): 1003–1004.

The idea that the formation of the Isthmus of Panama contributed to the formation of ice caps in the Arctic is in: https://www.whoi.edu/oceanus/feature/how-the-isthmus-of-panama-put-ice-in-the-arctic/.

For a discussion of how refugees from glacial ice contributed to the evolution of diverse species, and what that means for human evolution, see: J. R. Stewart and C. B. Stringer, "Human evolution out of Africa: the role of refugia and climate change," *Science* 335 (2012): 1317–1321.

Modeling the relationship between climatic change and the evolution and migration of *Homo heidelbergensis*, *Homo neanderthalensis*, and *Homo sapiens* is in: A. Timmermann et al., "Climate effects on archaic human habitats and species successions," *Nature* 604 (2022): 495–501 (the paper is open access at: https://www.nature.com/articles/s41586-022-04600-9. The lead author, Axel Timmermann, has an excellent online talk on the work: https://www.youtube.com/watch?v=8a8m5S6GmXc/.

A recording and transcript of David Foster Wallace's commencement speech "This Is Water" is at *https://fs.blog/david-foster-wallace-this-is-water* and in print: David Foster Wallace, *This Is Water: Some Thoughts, Delivered on a Significant Occasion, about Living a Compassionate Life* (New York: Little, Brown, 2009).

## CHAPTER 7: ON THIN ICE

Information on the hamlet of Grise Fiord, including information for visitors, is at: https://travelnunavut.ca/regions-of-nunavut/communities/grise-fiord/.

The Niels Bohr Institute of the University of Copenhagen has an excellent biographical site about the different phases of Willi Dansgaard's life and career: https://nbi.ku.dk/english/www/willi/dansgaard/.

Camp Century is discussed in: Kristian H. Nielsen and Henry Nielsen, *Camp Century: The Untold Story of America's Secret Arctic Military Base under the Greenland Ice* (New York: Columbia University Press, 2016).

Willi Dansgaard published numerous papers on oxygen isotopes and temperature records, including his analysis on the Camp Century core: W. Dansgaard, S. J. Johnsen, J. Møller, and C. C. Langway Jr., "One thousand centuries of climatic record from Camp Century on the Greenland ice sheet," *Science* 166 (1969): 377–381. Select other papers include: W. Dansgaard and S. J. Johnsen, "A flow model and a time scale for the ice core from Camp Century, Greenland," *Journal of Glaciology* 8 (2022): 215–223; W. Dansgaard et al., "Evidence for general instability of past climate from a 250-kyr ice-core record," *Nature* 364 (1993): 218–220; and W. Dansgaard et al., "A new Greenland deep ice core," *Science* 218 (1982): 1273–1277.

Dansgaard–Oeschger events are presented in: W. Dansgaard et al., "North Atlantic climatic oscillations revealed by deep Greenland ice cores," *Climate Processes and Climate Sensitivity* 29 (1984): 288–298.

Heinrich's classic paper on the rafting icebergs is: H. Heinrich, "Origin and consequences of cyclic ice rafting in the northeast Atlantic Ocean during the past 130,000 years," *Quaternary Research* 29 (1988): 142–152. Wally Broecker, an influential geochemist at Columbia University, proposed a mechanism that triggers Heinrich events: W. Broecker, G. Bond, M. Klas, E. Clark, and J. McManus, "Origin of the northern Atlantic's Heinrich events," *Climate Dynamics* 6 (1992): 265–273.

The 2002 International Panel on Climate Change quote that showed the classic view of a stable Antarctic Ice Sheet is on page 9 of this report: https://www.ipcc.ch/site/assets/uploads/2018/05/SYR_TAR_full_report.pdf.

John H. Mercer's papers on West Antarctica include: J. H. Mercer, "Glacial geology of the Reedy Glacier area, Antarctica," *Geological Society of America Bulletin* 79 (1968): 471–486; and J. H. Mercer, "West Antarctic Ice Sheet and $CO_2$ greenhouse effect: a threat of disaster," *Nature* 271 (1978): 321–325. Mercer's colleague Terry Hughes published a similar warning call in: T. J. Hughes, "The weak underbelly of the West Antarctic Ice Sheet," *Journal of Glaciology* 27 (1981): 518–525. A retrospective of Mercer and his prediction is at: https://www.nature.com/articles/d41586-018-01390-x/.

Elizabeth Kolbert and Jeff Goodell have published important long-form journalism on the situation in West Antarctica. Goodell, writing in *Rolling Stone*, famously labeled Thwaites Glacier "the Doomsday Glacier" in: https://www.rollingstone.com/politics/politics-features/doomsday-glacier-thwaites-antarctica-climate-crisis-1273841/. Kolbert's influential piece is: https://www.newyorker.com/tech/annals-of-technology/the-west-antarctic-ice-sheet-melt-defending-the-drama/.

Pedro Skvarca's analysis of satellite data of the Larsen Ice Sheet is in: P. Skvarca, "Fast recession of the northern Larsen Ice Shelf monitored by space images," *Annals of Glaciology* 17 (1993): 317–321.

Interviews with the Argentine scientists at Matienzo Station recall how terrifying the ice breakup was for those on the ice itself: https://www.scientificamerican.com/article/scientists-trek-collapsing-glaciers-antarctica-meltdown/. Details of the breakup are at: https://nsidc.org/learn/ask-scientist/what-happened-larsen-ice-shelf. NASA's time-lapse images of the breakup are at: https://earthobservatory.nasa.gov/world-of-change/LarsenB/.

Eric Rignot's satellite analyses of West Antarctica include: E. Rignot et al., "Four decades of Antarctic Ice Sheet mass balance from 1979–2017," *Proceedings of the National Academy of Sciences* 116 (2019): 1095–1103; E. Rignot et al., "Widespread, rapid grounding line retreat of Pine Island, Thwaites, Smith, and Kohler Glaciers, West Antarctica, from 1992 to 2011," *Geophysical Research Letters* 41 (2014): 3502–3509; E. Rignot et al., "Recent Antarctic ice mass loss from radar interferometry and regional climate modelling," *Nature Geoscience* 1 (2008): 106–110; and E. Rignot et al., "Acceleration of Pine Island and Thwaites Glaciers, West Antarctica," *Annals of Glaciology* 34 (2002): 189–194. The workings of the MODIS satellite imaging system is presented at: https://modis.gsfc.nasa.gov/.

Eric Rignot's inaugural address to the National Academy of Sciences is a short précis of what his satellite work means for understanding our future: https://www.youtube.com/watch?v=AT5jdqHUeKM/.

The paper that describes the shipboard submersible and its analysis of the seafloor under Thwaites Glacier is https://www.nature.com/articles/s41561-022

-01019-9 and is discussed for a general audience here: https://eos.org/articles/seafloor-reveals-a-period-of-rapid-retreat-for-thwaites-glacier/.

Two scientific papers on the Icefin robotic submersible are: P. E. Davis et al., "Suppressed basal melting in the eastern Thwaites Glacier grounding zone," *Nature* 614 (2023): 479–485; and B. E. Schmidt et al., "Heterogeneous melting near the Thwaites Glacier grounding line," *Nature* 614 (2023): 471–478.

Charles Francis Hall's disastrous expedition is chronicled in: Chauncey Loomis, *Weird and Tragic Shores: The Story of Charles Francis Hall, Explorer* (New York: Modern Library, 2000). Hall's travels among Inuit are in: *Arctic researches, and life among the Esquimaux: being the narrative of an expedition in search of Sir John Franklin, in the years 1860, 1861, and 1862,* freely available from the Internet Archive at: https://archive.org/details/arcticresearches00hall.

The way that the rising land, known as eustatic rebound, can affect West Antarctica and sea-level rise is in: V. Coulon et al., "Contrasting response of West and East Antarctic Ice Sheets to glacial isostatic adjustment," *Journal of Geophysical Research: Earth Surface* 126 (2021): e2020JF006003.

The octopus genome work and its relationship to West Antarctica is in: S. Lau et al., "Genomic evidence for West Antarctic Ice Sheet collapse during the last interglacial period," bioRxiv 2023.01.29.525778 (open-access preprint).

Jeff Goodell wrote a powerful book on how sea-level rise can change our world: *The Water Will Come: Rising Seas, Sinking Cities, and the Remaking of the Civilized World* (New York: Little, Brown, 2017).

The National Atmospheric and Oceanographic Administration's *Sea Level Rise Technical Report* for 2022 is at: https://oceanservice.noaa.gov/hazards/sealevelrise/sealevelrise-tech-report.html/.

The Atlantic meridional overturning circulation (AMOC) is the name of the current that brings heat and saltwater north. A recent paper makes the controversial prediction that this current will collapse with the increasing melting of glaciers: P. Ditlevsen and S. Ditlevsen, "Warning of a forthcoming collapse of the Atlantic meridional overturning circulation," *Nature Communications* 14 (2023): 1–12 (open access at: https://www.nature.com/articles/s41467-023-39810-w). The Science Media Centre published the conflicting opinions of diverse scientists on this result: https://www.sciencemediacentre.org/expert-reaction-to-paper-warning-of-a-collapse-of-the-atlantic-meridional-overturning-circulation/. The effects of a changing AMOC on the climate is at: https://news.mongabay.com/2023/11/climate-change-threatens-the-north-atlantics-currents-ecosystems-and-stability-commentary/.

The different predictions of sea-level change as a consequence of melting ice appeared in the journal *Nature* in 2021: R. M. DeConto et al., "The Paris Climate Agreement and future sea-level rise from Antarctica," *Nature* 593 (2021): 83–89, https://doi.org/10.1038/s41586-021-03427-0; and T. L. Edwards et al., "Projected land ice contributions to twenty-first-century sea level rise," *Nature* 593 (2021): 74–82, https://doi.org/10.1038/s41586-021-03302-y. Jeff Goodell had an informative account of both papers, including comments from researchers, in the May 21,

2021, issue of *Rolling Stone* titled "Antarctica's Doomsday Glacier: How Doomed Are We?"

## CHAPTER 8: HEATING UP

Aerial photos of the polar regions can be purchased online: https://natural-resources.canada.ca/maps-tools-and-publications/satellite-imagery-elevation-data-and-air-photos/air-photos/national-air-photo-library/9265/.

For information on the rhythms of the year for Inuit populations, see: https://www.cbc.ca/news/canada/north/names-months-inuktitut-aseena-mablick-1.3977403 and Richard G. Condon, *Inuit Behavior and Seasonal Change in the Canadian Arctic* (Ann Arbor: UMI Research Press, 1983).

The Nunavut Climate Change Secretariat website "Voices from the Land" has information on how climate change has affected diverse communities. It gathers quotes from people in different hamlets, including those in the text: https://climatechangenunavut.ca/en/knowledge/voices-land.

The article in the *Robb Report* on the North Pole cruise is: https://robbreport.com/motors/marine/north-pole-ice-breaking-ship-1234828830/.

Evidence suggests that, on average, the Arctic is heating four times faster than the rest of the planet: M. Rantanen et al., "The Arctic has warmed nearly four times faster than the globe since 1979," *Communications Earth & Environment* 3 (2022): 168–123 (open access at: https://www.nature.com/articles/s43247-022-00498-3).

NASA maintains information on the state of sea ice at https://climate.nasa.gov/vital-signs/arctic-sea-ice/, while NOAA maintains an "Arctic Report Card" on the state of the region: https://arctic.noaa.gov/report-card/.

The extreme warmth of recent summers in Siberia is documented in: R. M. Hantemirov et al., "Current Siberian heating is unprecedented during the past seven millennia," *Nature Communications* 13 (2022): 968 (open access at: https://www.nature.com/articles/s41467-022-32629-x). Reports of lightning in the Arctic are discussed in: https://www.theguardian.com/environment/2022/jan/07/lightning-high-arctic-rise-scientists-worried/.

*Canadian Geographic* has an account of the dispute between Canada and Denmark, known as the Whiskey War: https://canadiangeographic.ca/articles/canada-denmark-end-50-year-whiskey-war-over-hans-island/.

The remarkable bush plane that has made polar research possible is the Twin Otter, made by De Havilland, Canada. The planes we use in the Arctic summer move to the Antarctic for use by researchers in summer there. You can find accounts of the incredible things they can do at: https://airfactsjournal.com/2021/02/runways-optional-twin-otter-tales-from-the-arctic/.

The scientific results of Operation Franklin, including maps, is published in: Yves O. Fortier, *Geology of the North-central Part of the Arctic Archipelago, Northwest Territories (Operation Franklin)*, Vols. 1 and 2 (Ottawa: Department of Mines and Technical Surveys, 1963).

The founding of the Polar Continental Shelf Project (PCSP) is discussed here: https://science.gc.ca/site/science/en/educational-resources/history-geological-survey-canada-175-objects/79-polar-continental-shelf-project-1958. The history of PCSP, as well as Canadian polar science in general, is covered in the obituary of George Hobson, an influential director: https://csegrecorder.com/articles/view/in-memoriam-of-george-d-hobson/.

The story of Mary Dawson's remarkable career, as well as her scientific discoveries, can be found in: https://carnegiemuseums.org/carnegie-magazine/spring-2021/daring-to-dig/; and J. J. Eberle and M. C. McKenna, "The indefatigable Mary R. Dawson: Arctic pioneer," *Bulletin of Carnegie Museum of Natural History* 39 (2007): 7–16. Her encounter with the wolves is in: K. Munthe and J. H. Hutchison, "A wolf-human encounter on Ellesmere Island, Canada," *Journal of Mammalogy* 59 (1978): 876–878.

Background on the Lomonosov Ridge and the ways that science is being used in land claims are in: https://www.bbc.com/future/article/20200722-the-rush-to-claim-an-undersea-mountain-range. An analysis of the seabed at the ridge is in: C. Marcussen et al., "Age and origin of the Lomonosov Ridge: a key continental fragment in Arctic Ocean reconstructions," in *EGU General Assembly Conference Abstracts* (April 2015): 10207 (https://ui.adsabs.harvard.edu/abs/2015EGUGA..1710207M/abstract).

Details on the workings of the United Nations Committee on the Law of the Sea (UNCLOS) are at: https://www.un.org/depts/los/convention_agreements/texts/unclos/unclos_e.pdf/.

Russia's claim to the Arctic is discussed in: https://polarjournal.ch/en/2023/02/21/russias-claim-to-north-pole-territory-officially-confirmed/; https://www.theguardian.com/world/2007/aug/02/russia.arctic; and https://polarconnection.org/russia-extended-continental-shelf-arctic/.

The text of the Ilulissat Declaration is at: https://arcticportal.org/images/stories/pdf/Ilulissat-declaration.pdf/.

The history of Inuit land claims is at: https://www.adventurecanada.com/the-nunas-an-introduction-to-the-four-inuit-homelands-in-canada. The genetic history of the peoples of the Canadian Arctic is in: M. Raghavan et al., "The genetic prehistory of the New World Arctic," *Science* 345 (2014): 1255832. The Canadian Museum of Nature has an online exhibition on the Dorset culture at: https://www.historymuseum.ca/cmc/exhibitions/archeo/paleoesq/peexheng.html (the site includes an excellent bibliography).

Larry Audlaluk's experience is in his book, *What I Remember, What I Know: Life of a High Arctic Exile* (Toronto: Inhabit, 2021).

The Qikiqtani Truth Commission's work, and the apology from the government of Canada, is at: https://www.qtcommission.ca/en/about-us/the-commission/.

The situation in Newtok is discussed in: https://relocatenewtok.org/about/ and https://www.theatlantic.com/business/archive/2015/08/alaska-village-climate-change/402604/. The quotes in the text are taken from: https://5g.wilsoncenter.org/event/fleeing-change-relocating-the-village-newtok-alaska/.

The relocation settlement of Newtok and Napakiak is discussed in: https://

alaskapublic.org/2022/12/02/interior-department-puts-40m-toward-community-relocation-efforts-for-newtok-and-napakiak/; https://www.doi.gov/pressreleases/biden-harris-administration-makes-135-million-commitment-support-relocation-tribal; and https://alaskabeacon.com/2022/12/02/villages-chosen-for-relocation-grants-singled-out-for-climate-change-threats-and-progress/.

MIT's Climate Portal is an excellent web resource on permafrost and what melting permafrost means for your future: https://climate.mit.edu/explainers/permafrost. The Batagaika Crater, the massive hole in the permafrost of Siberia, can be seen from space at: https://earthobservatory.nasa.gov/images/90104/batagaika-crater-expands. An example of the ways that people are instructed to mitigate changing permafrost is at: https://dot.alaska.gov/traveltopics/building-in-permafrost.shtml/.

The subpopulations of polar bears, and their status, are described in: https://www.polarbearscanada.ca/en/polar-bears-canada/canadas-polar-bear-subpopulations; https://www.canada.ca/en/environment-climate-change/services/species-risk-public-registry/cosewic-assessments-status-reports/polar-bear-2018.html; and https://polarbearsinternational.org/polar-bears-changing-arctic/polar-bear-facts/status/.

Polar bear genetics, as well as polar bears' relationship with brown bears, is discussed in: C. Lindqvist et al., "Complete mitochondrial genome of a Pleistocene jawbone unveils the origin of polar bear," *Proceedings of the National Academy of Sciences* 107 (2010): 5053–5057; and T. Lan et al., "Insights into bear evolution from a Pleistocene polar bear genome," *Proceedings of the National Academy of Sciences* 119 (2022): e2200016119.

The situation with orcas and narwhals is analyzed in: G. A. Breed et al., "Sustained disruption of narwhal habitat use and behavior in the presence of Arctic killer whales," *Proceedings of the National Academy of Sciences* 114 (2017): 2628–2633. The ways in which the changing narwhal hunt affects Inuit life (as well as the source for the quote) are described in: https://www.nrdc.org/stories/melting-arctic-reckoning-save-indigenous-culture-narwhals-and-silent-sea. An article on the subject in the *Nunatsiaq News*, the major newspaper of Nunavut, is: https://nunatsiaq.com/stories/article/arctic-warming-may-bring-more-killer-whale-predation-on-nunavut-narwhals/.

## EPILOGUE

Glacial geoengineering is discussed in: J. C. Moore, R. Gladstone, T. Zwinger, and M. Wolovick, "Geoengineer polar glaciers to slow sea-level rise," *Nature* 555, no. 7696 (2018): 303–305; A. Lockley et al., "Glacier geoengineering to address sea-level rise: a geotechnical approach," *Advances in Climate Change Research* 11 (2020): 401–414; and J. C. Moore et al., "Targeted geoengineering: local interventions with global implications," *Global Policy* 12 (2008): 108–118.

The broken-wing display is discussed in: https://www.researchgate.net/profile/Miguel-GomezSerrano/publication/322209976_Broken_Wing_Display/links/5a670a8d0f7e9b76ea8d6163/Broken-Wing-Display.pdf/.

# ILLUSTRATION CREDITS

Cover modified from a photo courtesy of the author

iv–v: Photo courtesy of Sune Tamm showing Aztec Mountain Camp, 2,500 feet above the Taylor Glacier in the distance

x: Map by Chris Erichsen

xi: Map by Chris Erichsen

16: Photo courtesy of Robert Schwartz

18: Photo courtesy of Shridhar Anandakrishnan, Pennsylvania State University

24: Photo from www.pixabay.com is in the public domain.

28: Photo courtesy of Byron Adams, Brigham Young University

30: Photo courtesy of the author

32: Photo courtesy of Eric Rignot

48: Photo courtesy of Byron Adams, Brigham Young University

57: Photo courtesy of Ted Daeschler, Academy of Natural Sciences of Drexel University

59: Photo by Mike Beauregard via creative commons license is in the public domain.

68: Photo courtesy of Ted Daeschler, Academy of Natural Sciences of Drexel University

84: Photo by Xavier Mertz from the State Library of New South Wales is in the public domain.

90: Photo from Masaru Yoshida, from "Discovery of the Yamato Meteorites in 1969," *Polar Science* 3 (2010): 272–284

93: Photo from NASA is in the public domain.

97: Photo from NASA/Cindy Evans is in the public domain.

# ILLUSTRATION CREDITS

109: Photo by Mike Lucibella, *The Antarctic Sun*, January 2017
118: Photo from National Library of Norway is in the public domain.
120: Image from New York Public Library Digital Collection is in the public domain.
123: Photo of engraving by J. M. Butler based on a sketch by Elisha Kent Kane, published in Kane, *Arctic Explorations* (1856), is in the public domain.
129: Photo permission from Scott Polar Research Institute, University of Cambridge
130: Photo permission from Scott Polar Research Institute, University of Cambridge
132: Photo of *Glossopteris* from Houston Museum of Natural History is in the public domain.
145: Photo courtesy of Stephen Gatesy, Brown University
153: Photo courtesy of Maureen Raymo, Columbia University in the City of New York
167: Photo courtesy of the Niels Bohr Institute of the University of Copenhagen
169: Photo from U.S. Army Corps of Engineers is in the public domain
173: Photo and permission from the Ohio State University Archives, John H. Mercer Papers, SPEC.PA.56.0024, Box 6, folder 3, image 42
180: Photo from NASA is in the public domain.
186: Photo courtesy of Rob Robbins
203: Photo courtesy of Anthony Jarvis, Buffalo Aviation
208: Photo courtesy of Jaelyn Eberle, The University of Colorado, Boulder
213: Map by Chris Erichsen
219: Photo by permission from Matisse Harvey/Radio-Canada

# INDEX

Note: Italicized page numbers indicate material in photographs or illustrations.

# INDEX

# INDEX

# INDEX

# INDEX

## INDEX

# INDEX

## ABOUT THE AUTHOR

**Neil Shubin** is the bestselling author of *Your Inner Fish*, *The Universe Within*, and *Some Assembly Required*. He is the Robert R. Bensley Professor of Organismal Biology and Anatomy at the University of Chicago. He was the host of the Emmy Award–winning PBS miniseries *Your Inner Fish*, which was based on his bestselling book. He was elected to the National Academy of Sciences in 2011. Educated at Columbia, Harvard, and the University of California, Berkeley, he lives in Chicago.